Build your own
DRONE
Manual

COVER IMAGES:

Upper, and lower left and right – *Author*
Lower centre – *Rolando Cerna*

© Alex Elliott 2016

All rights reserved. No part of this publication may be reproduced or stored in a retrieval system or transmitted, in any form or by any means, electronic, mechanical, photocopying, recording or otherwise, without prior permission in writing from Haynes Publishing.

First published in February 2016

A catalogue record for this book is available from the British Library.

ISBN 978 0 85733 813 6

Library of Congress control no. 2015948104

Published by Haynes Publishing,
Sparkford, Yeovil,
Somerset BA22 7JJ, UK.
Tel: 01963 440635
Int. tel: +44 1963 440635
Website: www.haynes.co.uk

Haynes North America Inc.,
861 Lawrence Drive, Newbury Park,
California 91320, USA.

Printed in the USA by
Odcombe Press LP,
1299 Bridgestone Parkway,
La Vergne, TN 37086.

Build your own DRONE Manual

Owners' Workshop Manual

The practical guide to safely building, operating and maintaining an Unmanned Aerial Vehicle (UAV)

Alex Elliott

Contents

6	A brief history of drones
What is a drone?	8
History of drones	8

14	Types of drones
Scratch-build or off-the-shelf?	16
Fixed-wing	16
Rotorcraft	19

26	Applications of drones
Mapping and inspection	28
Aerial photography	31
Surveillance	32
Delivery by drone	33
Hobbyists	34

36	Anatomy of a drone
Airframe	38
Autopilot	42
Radio control	52
Motors	57
Electronic speed controllers	59
Propellers	66
Batteries	71
Gimbals and cameras	74
First-person view	80
Choosing components for your drone	90

96	Drone builds
Aerial photography drone	98
Mini FPV quadcopter	116
Fixed-wing drone	124

140	Flying and safety
Safety information	142
Learning to fly	144
Pre-flight safety checks	145
Where to get help	149

150	Appendix
Thrust data tables	150

152	List of abbreviations

153	Index

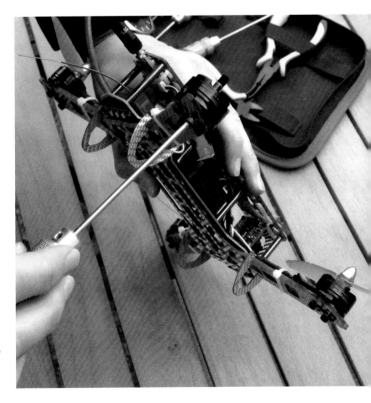

OPPOSITE An S550 hexacopter carrying a GoPro camera on a stabilised gimbal to shoot aerial videos. Note the GPS/Compass module mounted on a mast above the other electronics to ensure that it is free from interference. *(Author)*

RIGHT Securing the motors on the Silver Blade FPV quadcopter. It's always a good idea to use thread lock on all the screws to ensure they don't work loose due to vibration. *(Sam Evans)*

CHAPTER 1
A BRIEF HISTORY OF DRONES

(Pixabay)

What is a drone?

The term 'drone' is often given bad press. In the media it's linked with military air strikes and warfare. However, drones are much more than weapons. They're used by thousands of hobbyists and organisations around the world, for entertainment, research, conservation, agriculture and even to save lives. Consequently we need to reclaim the term drones so that people start to understand their full potential.

In terms of the technicalities, many people define drones or 'UAVs' – unmanned aerial vehicles – as having the ability to fly entirely autonomously from take-off to landing. In which case some of the quadcopters for sale today aren't technically drones at all, but rather remote-controlled quadcopters. Although they do have some degree of autonomy they can't fly entirely on their own from automatic take-off, visit a series of waypoints and then land unaided. Similarly, many of the so-called drones used by the military are actually remotely piloted vehicles that always have someone flying them.

In this book, I use the term 'drone' quite loosely to encompass all radio-control aircraft that have some sort of flight controller.

History of drones

In order to fully appreciate the history of drones we need to look at the broad history of unmanned aircraft, which has seen parallel developments alongside manned aviation from the very beginning of human flight. They have a dark history, as they've often been used as powerful weapons, and still are today. However, recent civilian applications have seen them being regularly used in positive and lifesaving roles.

Depending on how you want to define the word 'drone', one could argue that unmanned flying contraptions were used as early as 1849, when balloons armed with bombs were used by the Austrian army during an attack on Venice. However, sticking to more conventional wisdom the first flight of an unmanned aircraft can be dated back to the 1890s, when Otto Lilienthal, a German aviation pioneer, experimented with unmanned gliders to test lightweight lifting-wing designs. As with many early designs of highly experimental aircraft, these were unmanned to avoid injuring brave test pilots. As a result Lilienthal was able to test many of his bold designs safely and learn from the crashes.

At around the same time, in 1896 Samuel

RIGHT Samuel P. Langley's Aerodrome Number 6 made the world's first successful long-range flight of an unpiloted, engine-driven, heavier-than-air craft. It was launched by catapult from the top of a houseboat on the Potomac River, Virginia, USA, in November 1896, and flew for almost a mile. It was powered by a single-cylinder steam engine driving two pusher propellers through a geared transmission. *(Wikipedia)*

ABOVE **Full-size Kettering Bug model on display at the National Museum of the United States Air Force in Dayton, Ohio. The main undercarriage detached from the aircraft during take-off.** *(Greg Hume)*

Pierpont Langley was experimenting with steam-powered aircraft. Interestingly his aircraft were launched using a catapult system, a method still used by many modern drones. Langley's unmanned, unguided aircraft known as the 'Aerodrome' managed to fly just short of a mile along the Potomac River as part of the testing process for a proposed manned plane. Though the project was ultimately abandoned, the test flight was significant in the world of aviation as it was the first successful, long-range flight of a self-propelled aircraft. This was several years before the Wright brothers' famous first flight.

Despite the successful flights of unmanned test aircraft, it was quickly realised that having a pilot controlling the plane was more useful in order to progress design and development. Having pilots aboard was therefore a risk that needed to be taken. The Wright brothers' historical flight taught the world that controlled flight could be achieved by using wing warping to manage aircraft roll, and this breakthrough led to a boom in technical advancements for the aviation industry. The success of the Wright brothers was the catalyst that shaped aviation science into what it is today.

Not long after, Lawrence Sperry – known as 'the grandfather of autopilots' – made use of the gyroscopes his family company invented to construct the first autopilots to help pilots control their planes. In 1914 Sperry's autopilot enabled a manned aircraft to fly straight and level, greatly reducing the pilot's workload.

It was also around this time that Professor Archibald Low played an important role in developing radio guidance systems to control aircraft remotely, culminating in the remotely controlled flight of the Ruston Proctor AT (aerial target) in 1917. It should also be noted that inventor Nikola Tesla had already demonstrated

ABOVE **Helicopter designed by George De Bothezat, descending at McCook Field after remaining airborne for 2 minutes 45 seconds.** *(Wikipedia)*

the ability to remotely control a boat many years earlier, in 1898.

During World War One the Kettering Bug programme was the first major project to employ unmanned aircraft. This biplane, which was essentially a flying torpedo, had a guidance system designed by Elmer Sperry, the father of Lawrence, and was powered by a reciprocating engine that used a dolly and rail arrangement for take-off. In theory the Kettering Bug could autonomously navigate up to 40 miles (64km) to its target, guided by a gyroscope-based autopilot with an aneroid barometer to monitor altitude. In order to measure the distance flown the aircraft used a mechanical system based on the number of engine revolutions required for the aircraft to reach its target. However, the Kettering Bug was never deployed because World War One ended before it could be completed.

It was in 1922 that the first successful quadcopter-style aircraft was successfully flown. Known as the De Bothezat helicopter, it used rotors in an X-shaped structure. Around 100 flights were completed, with the highest altitude attained being 5m. However, the design was never taken any further due to its mechanical complexity and the exceptionally high pilot workload required during a hover.

Between the world wars unmanned aircraft saw limited development in terms of autonomy, but advances in radio frequency (RF) transmissions made it easier to pilot aircraft remotely, rendering autopilots unnecessary. Various minor improvements were also made to autopilots and actuators. The main use of such systems was as aerial targets for gunnery practice, many of which were produced by British companies. Radio-controlled targets for pilot training saw prolific use of unmanned aircraft during this period, with some 12,000 being produced. Even today this remains a major military application for drones, although they now have a much greater level of autonomy.

World War Two saw another boom in the advancement of unmanned aircraft, particularly in Germany. Among the many air-to-ground or cruise missiles developed by Germany during this period, the most well-known was the Fieseler V-1, commonly known as the 'Buzz Bomb' due to the characteristic sound produced by its pulsejet engine. The Luftwaffe also used radio-controlled glider bombs to attack warships, the most common being known as the Fritz X. These were used by the Luftwaffe during the sinking of the *Roma*, an Italian warship, as it sailed to surrender to the Allies. Interestingly, the *Roma*'s wreckage was discovered only a few years ago in 2012 by another drone, albeit an underwater robot this time.

The Allies also utilised unmanned aircraft during World War Two. One such example was the TDR-1 assault drone used by the US Navy. A modified twin engine Interstate, this was probably the first video-piloted missile, being remotely controlled by an operator flying in a chase plane who could see what the TDR-1 could see via an on-board camera that broadcast a signal back to him.

Another US development programme, going by the name of Project Aphrodite, was commissioned to convert bombers such as the famous B-17 into unmanned aircraft. Again the concept of video piloting was used in which the aircraft was remotely guided via radio control, the remote pilot being able to see what the aircraft could see and read the cockpit instruments via a remote video feed from on-board cameras. The aircraft would take off manually with a crew aboard, but once airborne the crew would arm the bombs, hand over control to the remote pilot and parachute out while still over friendly territory. The aircraft would then continue on to its target area under remote control. Typically these missions were carried out on underground targets deep in enemy territory, such as V-1 bomb factories. However, the project proved unsuccessful due to its complexity and expense, and also because ground-penetration bombs developed by the British enjoyed greater success in such missions. Later these remotely controlled unmanned aircraft were used during nuclear bomb testing, to fly through the resultant mushroom cloud to study the radiation effects via specialist on-board sensors.

After World War Two drones were mainly used for target practice during pilot training. There were a few cases of converting manned aircraft into flying bombs during the Korean War, but there were no groundbreaking advances during this period due mainly to the focus of development being on cruise missiles, largely based on German research on the Fieseler V-1. This eventually led to the cruise missiles we have today which are capable of flying autonomously to their target.

The Ryan Aeronautics Company in California was the leader in aerial target drones in the 1960s, and the most famous unmanned drones of the period were the Teledyne Ryan Firefly/Firebee family, which have been amongst the most widely produced target drones to

BELOW V-1 cutaway, showing some of the main components. *(USAF)*

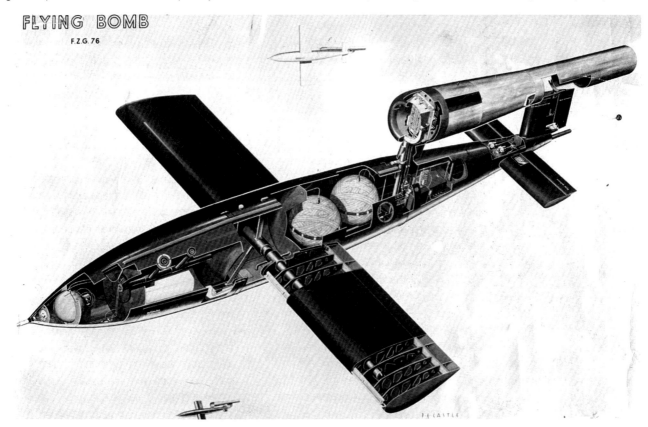

ABOVE Lockheed DC-130 drone control aircraft carrying two BQM-345 Firebee target drones under its wing. *(Wikipedia)*

date. Due to their success with target drones the company was also asked to develop a surveillance variant, which first flew in 1964. Some versions of the Firefly saw service in Vietnam on surveillance and reconnaissance missions, such as battle-damage assessment. With over 7,000 of all variants eventually being built, Firefly/Firebee drones are undoubtedly the forerunners of all modern drones. Many still serve in militaries around the world and some were used as recently as 2003 during the Iraq invasion, where they flew ahead of piloted planes to lay down anti-missile chaff corridors.

As the years went by and technology improved drones became a more reliable solution for reconnaissance operations, leading to the widespread use of the infamous Predator drone amongst others. Due to its ability to fly for long periods at great height, this drone started out as a purely reconnaissance platform, but armed versions began to appear around 2001 and continue in use today.

Dawn of hobby drones

As with most military technology, it was only a matter of time before drones began to be adopted for civilian use. The beginnings of small, low-cost hobby drones can be linked to the Paparazzi UAV project that started in 2003. The Paparazzi project's aim was to develop an open source autopilot system, and it has enjoyed many iterations over the years and is still under active development today. Its open source nature has greatly aided the drone hobby, as the developers publish all the source code and autopilot design files. These are used in the drone community to add new features, fix bugs and improve hardware designs. Despite the efforts of the early Paparazzi developers they were limited by the lack of powerful and cheap sensors, and because their early work on the autopilot was merely to achieve flight, little effort was put into making the system user-friendly. This in turn limited the adoption of hobby drones by universities and their students

ABOVE **There's a large selection of autopilot boards for you to choose from.** (Author)

as well as other enthusiasts who already had a fair amount of electronics knowledge.

It was only around 2009 that sensors and processors started to become very cheap, thanks to smartphone technology. Because of the mass-production nature of smartphones the technology quickly started to improve and get even cheaper, so that an inertial sensor that would once have cost hundreds of pounds now cost just a few, due to economies of scale. This meant that developers could now integrate better sensors into their autopilot projects for much less money. In fact most of today's hobby autopilots essentially have many of the same sensors as the average smartphone. There are even several projects that have turned an actual smartphone into an autopilot!

It was also around 2009 that the ArduPilot open source autopilot project was started, with the aim of making drones more accessible to the masses. Initially based on the Arduino electronics platform, their first autopilot used thermopiles to keep the aircraft stable. Thermopiles measure the amount of thermal energy, and by using several (facing upwards, forwards, downwards, left and right) the angle of the aircraft can be estimated based on the difference in temperature between the sky and the ground. Newer versions of the ArduPilot subsequently started to use inertial sensors like accelerometers and gyroscopes, as they offer far superior accuracy.

Each year the ArduPilot project gained momentum, added new features and made things even more user-friendly. Today ArduPilot software is able to control all unmanned vehicles, be they fixed-wing aircraft, multirotors, boats or ground vehicles. Several other autopilot projects are also around, with some focusing on specific platforms, but the ArduPilot project is probably the most popular drone platform in use, allowing you to turn your R/C aircraft into a drone capable of flying on its own from take-off to landing.

CHAPTER 2
TYPES OF DRONES

(David Fant)

OPENING SPREAD An example of a conventional configuration equipped with an FPV video system enabling the pilot to receive real-time video from the aircraft while it's flying. This aircraft has the motor mounted in a pusher configuration so the propeller isn't in view of the camera.
(David Fant)

Drones are usually categorised by either configuration or size. The configuration can be broken down into two main types: fixed-wing, which is just like traditional aeroplanes, where lift is created by moving the wings forward through the air; and rotorcraft, in which the wings (rotors) spin around to provide lift, as on a helicopter. Drones are also classified by total size/mass. In this book we'll only focus on small to micro drones that are below 7kg total mass, as this is currently the maximum weight that doesn't have too many regulations. It's very uncommon for a hobby drone to weigh more than 7kg, as drones that are heavier often involve specialised equipment intended for commercial use.

Scratch-build or off-the-shelf?

You might come across many scratch-build aircraft – both multirotor and fixed-wing – that pilots have designed and built themselves. However, although building an aircraft from scratch can be a very rewarding experience once you get it right, it can also be expensive and frustrating if you don't know what you're doing. So my suggestion would be that you start with an off-the-shelf aircraft frame first so that you can focus on all the other aspects of building one, such as configuring your autopilot and matching your first-person-view (FPV) gear, etc. Adding to the equation a custom-built frame that may or may not fly correctly just creates an extra complication. Nevertheless, custom-building your own aircraft is something that you should definitely attempt later on as you become more familiar with R/C aircraft and drones in general.

If you'd like to go down the path of designing a scratch-build, fixed-wing drone you should spend some time learning about the basics of aerodynamics first, to ensure that you design something that will fly. Alternatively there are countless plans you can find online, which you can follow or use as a basis for your own design. Full drone kits are available too that you only need to assemble, which is a great way to learn the basics of drone building. Another advantage of building your own drone – apart from the financial savings – is that you'll understand how it works, which makes things easier when it comes to repairs or upgrades.

However, if you just want to focus on flying you can always buy a ready-to-fly drone, which arrives fully assembled and set-up so that you only need to charge up the batteries and you're ready to go. Ready-to-fly drones are certainly a great way to get into the hobby, but after you've learnt to fly your next step should be to try your hand at building your own.

In terms of scratch-build aircraft, multicopter frames are easier to design and build than fixed-wing aircraft, as you don't need to worry about aerodynamics and other stability factors. Multirotor frames are just a bunch of arms that hold the motors, with a central section to hold your equipment. In this book we'll focus on building your own drone based on combining various components and integrating them with an off-the-shelf R/C aircraft platform.

Fixed-wing

Without going into too many technical details, fixed-wing aircraft use a source of propulsion (usually an electric motor and propeller in our case) to generate lift on the wings to enable the aircraft to fly. There are numerous configurations for fixed-wing aircraft that depend on any combination of factors, including wing placement, number of wings, main body style, control surfaces and tail style (or empennage, if you want to sound fancy). Rather than go into details of all the possible configurations, in this section I'll only cover the main configurations best suited to hobby drones and first-person-view flying. The subject of fixed-wing aircraft building would be an entire book in its own right and there are already many available, as well as lots of information online.

When an aircraft is flying, its angle (flight attitude) can change along three axes: the pitch axis, which involves the aircraft tilting upwards and downwards; the yaw axis, which involves the aircraft changing its heading; and the roll axis, which involves the aircraft rolling left and right.

In order to control a fixed-wing aircraft, the wings and tail section have some small flap-like areas called control surfaces. There are three primary control surfaces: ailerons that control

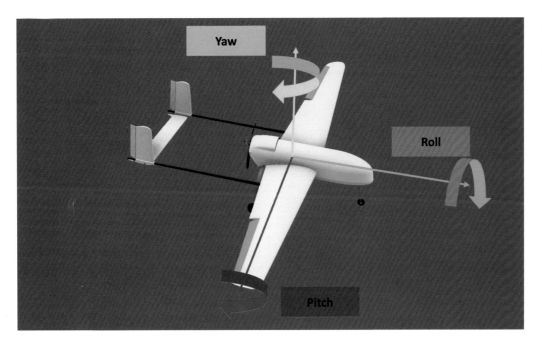

LEFT The three axes of motion of an aircraft. *(Author)*

the roll of the aircraft, a rudder that controls the yaw, and elevators that control the pitch.

Pusher or puller?

Irrespective of the wing arrangement of a fixed-wing drone aircraft, another design feature is classified by the location of the motor. Most of the time the motors are located at the rear of the aircraft, and are therefore classified as 'pusher' configurations, where the motor pushes the aircraft forward. Pusher motor configurations are quite popular for drone use because they keep the propeller out of view of on-board cameras when looking forward.

'Puller' (or 'tractor') configurations have their motor mounted with the propeller in the front, from where it pulls the aircraft forward. This is a more conventional configuration that can be slightly more efficient than a pusher, as the airflow hitting the propeller hasn't been disturbed by the aircraft's body. However, because the propeller is usually mounted in front of any camera that's being carried it can be annoying, due to the resultant 'rolling shutter' effect that shows up in its pictures as a strange linear pattern. However, for downward-facing cameras (as used for mapping UAVs) this isn't really an issue.

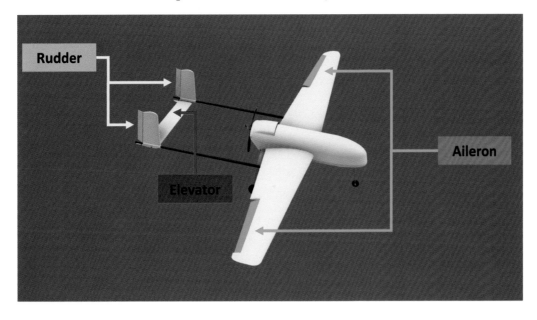

LEFT The three primary control surfaces on a fixed-wing aircraft. *(Author)*

ABOVE An electric foam aircraft commonly used as a trainer due to its stable flight characteristics. These trainer or Cessna-style planes aren't ideal for FPV since the propeller is mounted at the front, where it gets in the way of the forward-looking camera.
(Carsten Frenzl)

Conventional
Conventional configurations include a single-motor location either at the front of the aircraft (puller) or at the back (pusher). Large central wings are followed by a tail section.

Flying wing
Flying wings are fairly popular drone platforms as they have fewer components and are easier to transport. They're also more efficient and can cut through wind more easily than other drone platforms. A flying wing only has two control surfaces so only requires two servos, compared to more conventional aircraft that typically require four. The control surfaces on flying wings are called elevons, as they combine the functions of both an elevator (to pitch the aircraft up and down) and an aileron (to roll the aircraft left and right). An on-board flight controller takes care of knowing how much to move each control surface to perform the desired action.

Flying wings are great for drones, as there's lots of room to mount all your equipment and the motors and propellers are placed towards the rear, keeping them out of view of your camera. Their main drawback, however, is that the centre of gravity is more sensitive than on other configurations, and they can be slightly harder to fly if you aren't using an autopilot. You're also sometimes limited in the positioning of your equipment, since flying wing aircraft don't have as much space forwards and backwards for you to move things around to get the centre of gravity in its optimum position.

Twin-boom
Twin-boom aircraft have their tail section connected to the main body (known as the fuselage) via two booms. This allows for a larger propeller to be attached at the back. Just as with flying wings, pusher motors are ideal for use on twin-boom drones as the propeller doesn't get in the way of on-board forward-facing cameras. Also, since the twin-boom configuration still has a conventional tail section it's naturally more stable and easier to fly than a flying-wing design.

RIGHT A flying-wing drone passing overhead. As you can see, flying-wing platforms usually have just two control surfaces. These are called 'elevons', as they combine the function of elevators (to pitch up and down) and ailerons (to roll left and right).
(James Whomsley)

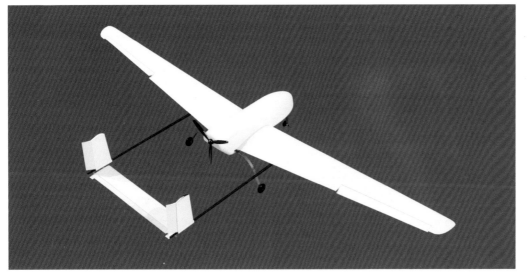

LEFT A 3D rendering of a twin-boom drone. The tail section is supported by two booms connected to the wing, which are spaced far enough apart not to interfere with the pusher propeller. *(Author)*

There are several variations of twin-boom aircraft, the principal of which has an inverted-V tail arrangement, which has the advantage of only having two control surfaces (opposed to three on a regular twin-boom) so that you only need to use two servos. This inverted-V tail configuration is also more efficient aerodynamically.

Rotorcraft

The term rotorcraft is used to categorise aircraft that achieve flight through spinning a rotor through the air to create enough lift to enable them to fly. Rotorcraft can be divided into two main categories, the first being multicopters and the other being conventional helicopters. Multicopters are classified according to the number of motors they use – for example, a quadcopter has four motors and a hexacopter has six motors. Each type is described in more detail below.

Conventional helicopter

A conventional helicopter is able to achieve control by changing the pitch and angle of its rotor blades. This is done via a complex mechanism known as a swashplate. Most helicopters usually have a single large rotor, which creates torque in the opposite direction to its rotation. This is why a tail rotor is needed to offset the yaw motion, in order to keep the helicopter facing the right way. This is consequently seen as a slightly inefficient design since some of the energy is used purely to keep the helicopter facing in the same direction rather than contributing to generating lift. To compensate for this, some helicopters have been designed with two counter-rotating blades, each spinning in opposite directions to neutralise the torque difference while using all the energy to generate lift.

Conventional helicopters aren't too common amongst hobby drones, mainly due to the cost and complexity of building them. They're also considerably more dangerous than other drones due to their much larger and heavier rotor blades. In addition their complexity means they're usually more fragile in a crash than common drone platforms such as quadcopters, which is another factor in the popularity of the latter.

BELOW An example of a typical conventional radio-control helicopter. *(Mike Lehmann)*

19
TYPES OF DRONES

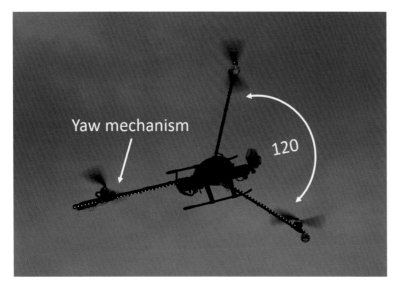

ABOVE Example of a tricopter. *(Martin Taraldset)*

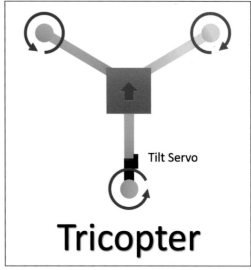

ABOVE RIGHT Top view of a tricopter configuration. The rear motor is mounted on a tilting mechanism that's able to adjust its angle via a servo. *(Author)*

Although I'd recommend that you stay away from a conventional helicopter as your first drone platform, one of the factors in their favour is how stable they are in windy conditions compared to multirotors. Also, if you need to build a long-range drone the availability of internal combustion R/C helicopters, which can fly for a lot longer than electric models, makes them a great choice.

Tricopter

Typically the naming convention for multirotor copters follows the Latin numbering system; so 'tri' denotes three motors, 'quad' denotes four, 'hexa' denotes six and so on, followed by the suffix '-copter'. Tricopters therefore use three motors, arranged in a triangular layout with one at the back and two up front. The arms of a tricopter are typically separated by 120°, which is an advantage when using an on-board camera since this wide separation angle means that the arms and propellers stay out of the camera's view.

Another benefit is that since tricopters only use three motors they're cheaper to build, though in order to fully control the tricopter the rear motor needs to tilt sideways to rotate the copter left and right (yaw). This also means that tricopters can yaw much faster than other configurations such as quadcopters. However, tricopters are typically more complex to build due to the yaw mechanism required to tilt the

RIGHT Close-up view of a yaw mechanism used to tilt the rear motor. *(Author)*

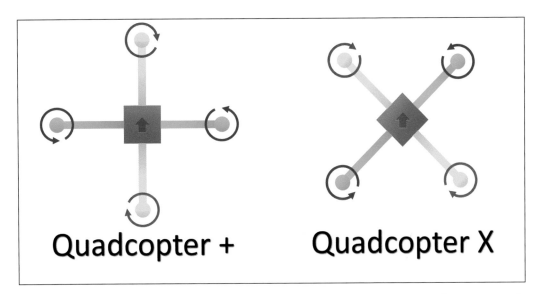

LEFT An example of the two types of quadcopter arrangement: on the left a plus (+) arrangement; on the right an ex (X) arrangement. *(Author)*

rear motor, which at least has the advantage of guaranteeing smoother flight characteristics that are beneficial when using your drone for filming.

Quadcopter

Quadcopters, which are the most common type of multirotor drone, consist of four motors arranged symmetrically in a plus (+) or ex (X) arrangement. In a plus arrangement the front of the quadcopter is aligned directly with a motor. In an ex arrangement the front of the quadcopter is directed between the front two motors. The latter configuration is more useful and common as it means the arms and propellers are less in view when using a forward-facing on-board camera.

Of the four motors on a quadcopter, two spin clockwise and the other two spin anticlockwise. This is to offset the torque created by the motors and keep the quadcopter facing in the right direction (it thus serves the same function as the tail rotor on a traditional helicopter). To control the quadcopter the speeds of the motors are varied – for example, to tilt the quadcopter forward the front two motors slow down while the rear two motors speed up.

The main reason for the popularity of quadcopters is because they're easy to build and control. To build one you only need to create a cross-frame with the four motors mounted on the ends of the arms, with no fancy mechanisms or linkages.

BELOW Example of a home-built quadcopter. *(Carsten Frenzl)*

ABOVE **Hexacopter with a stabilised GoPro gimbal. Notice that although the hexacopter is tilting the camera stays level.**
(Author)

Hexacopter

Hexacopters have – you guessed it – six motors. Apart from the obvious fact that having more motors means they can lift more equipment, an added benefit is that because the motors are spaced more closely around the centre if one motor should fail the hexacopter can usually stay relatively stable using the rest, thus allowing you to land. If a motor fails on a quadcopter or a tricopter it usually ends badly, as each motor is critical in keeping the drone stable. For this reason you'll see that more professional aerial photography drones usually have hexacopter or octocopter configurations, due to the extra payload and the precautionary redundancy of more motors.

Y6

The Y6 configuration is a mix of a tricopter and hexacopter. It has six motors on three arms arranged just like a tricopter, with two arms facing forward spaced at 120° and a single arm facing backwards. However, since Y6 copters have six motors in total they're still considered hexacopters. Each arm has two motors mounted on it, one facing upwards and one facing downwards. This is known as a coaxial motor arrangement. Typically each motor will spin in an opposing direction.

Y6 configurations have a few advantages over regular hexacopters. Since they have only three arms assembly is slightly easier and the frames can be slightly lighter. However, the

RIGHT **Top views of hexacopter configurations: on the left a plus (+) configuration, where the front of the drone is in line with the front arm; on the right an ex (X) configuration, where the front of the drone is between the front two arms.**
(Author)

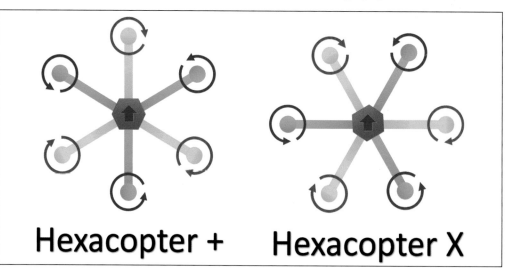

main advantage of this configuration is that the motors have more redundancy because of the way they're mounted, each motor pair acting on the same thrust axis. Should one motor fail the Y6 configuration will hardly notice any difference other than the one-sixth drop in thrust.

The one slight disadvantage of Y6 frames is that there's a marginal loss in efficiency (only about 5% or so) due to the bottom motors working in turbulent air that's been blown down from the top motors. However, this efficiency reduction is offset by the lighter weight of the frame (only three arms opposed to six), so you could consider this effect negligible. Sometimes you'll also find that the top propeller is slightly smaller than the bottom one.

Octocopter

Unsurprisingly, octocopters have eight motors, spaced evenly around them. These are amongst the biggest multicopters, with a typical aircraft having a 1m diameter. Just as with hexacopters, the increase in the number of motors means that octocopters can carry heavier payloads and have extra motor redundancy. While a hexacopter can generally only survive the failure of one motor (or possibly two if they're opposite one another), an octocopter can cope with more motor failures without crashing – depending on the payload and which motors fail.

Octocopters are commonly used as professional filming drones because of their payload capacity and motor redundancy leeway, as some film-grade cameras cost more than

ABOVE Example of a coaxial motor configuration on a Y6 hexacopter. The top motor spins clockwise while the bottom motor spins anticlockwise. *(Author)*

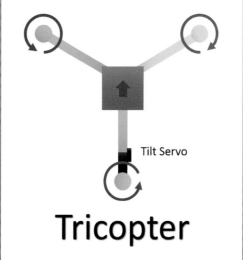

LEFT Top view of a Y6 hexacopter configuration. Here there are two motors on each arm, spinning in opposite directions. However, overall there are three clockwise motors and three anticlockwise to maintain symmetry and balance. *(Author)*

LEFT A Y6 hexacopter drone kit. Y6 frames look like tricopters but have six motors. *(Author)*

ABOVE An octocopter with DSLR camera, used for aerial filming. *(Robert Elzey)*

most cars and are very heavy. At the moment it's not very common to see octocopters in a hobby role due to their higher cost, but as time passes and the prices drop we should see more of them appearing.

X8

The X8 configuration is essentially a quadcopter frame with eight motors, a blend of a quadcopter and octocopter. X8 frames have four arms with two motors on each, one facing upwards and the other facing downwards. X8 copters have all the same advantages as octocopters, the main one being their ability to lift heavy payloads. Just like the Y6 configuration, the X8 also has the advantage of having two motors along the same thrust axis, so it's even more stable should a motor fail during flight (assuming the remaining seven produce enough thrust to keep the drone in the air).

X-copters
SingleCopter

Although x-copter isn't an official classification it's included here as there are many uncommon cases of copters that have a single motor (single copter/ducted fan). These use a single rotating propeller to produce thrust and are controlled via small flaps under the propeller that manipulate the airflow to allow the drone to rotate or pitch forward etc. These designs can be considered as very similar to conventional helicopters as they also use one main rotor to provide lift.

The advantage of a ducted fan copter is that they're mechanically simpler, as the blades of the main rotor are fixed and the small flaps underneath the blade are used to control the drone. Furthermore, because the blade is also

RIGHT Top view of an X8 octocopter. Here each arm has two motors mounted one above another, one spinning clockwise and the other spinning anticlockwise. *(Author)*

enclosed by the duct the propellers are slightly more efficient compared to rotor blades of the same size in a helicopter (as the duct reduces thrust losses at the tips of the blades for aerodynamic reasons that I won't go into).

Another important aspect of the duct in terms of safety is that it prevents the fast-spinning and dangerous blades from hurting anyone, and guards against debris getting caught up. However, this duct also has some drawbacks, the main one being wind stability. Because of their large frontal area single-copter drones are very susceptible to being pushed around by gusts of wind, making them unsuitable when the wind is very strong. Also, because the engines/blades are in the middle of the drone this limits the available space for mounting equipment.

Bicopter

Bicopters have two motors, and although they aren't particularly useful due to stability issues they do make for some cool-looking drones (think of the Pandora Warrior aircraft from the movie *Avatar*). Bicopters are more useful if you plan to build a hybrid drone (tiltrotor) that will transform into a forward-flying fixed-wing aircraft like the Boeing V22 Osprey, which can take off, hover and land just like a helicopter, but to fly longer distances tilts its rotors forward and transitions into a fixed-wing aircraft.

Due to their complexity it's not very common to see drones like these in our hobby. They're currently used mostly in commercial and military capacities. However, there are numerous developers and makers working on their own creations that are able to transform as described above, and such bicopters should become more 'mainstream' in the near future. On a side note, a tiltrotor drone isn't limited to just two motors.

More than eight motors

Sometimes you'll find, as a drone builder, that you can never have enough motors, so your imagination, budget and design skills are the only limits. I've seen some creations with as many as 18 motors! In reality you don't see too many multicopters with more than eight motors due to practicality and cost. However, there are some special situations where you need the extra lift from many motors but are limited by the maximum propeller/motor size, or it's just a fun and interesting project. Most of the time, for reasons of symmetry, an even number of motors is used, but there are always unique and cool creations that defy the conventions.

ABOVE A heavy-lift X8 octocopter. This has eight motors driving 15in propellers and a maximum take-off weight of 10kg, making it ideal to carry professional camera gear. *(Author)*

BELOW Example of a Honeywell RQ16, which is a single copter (ducted fan) drone. *(US Navy)*

CHAPTER 3
APPLICATIONS OF DRONES

(Rolando Cerna)

**OPENING SPREAD
A small quadcopter streams real-time videos back to its pilot as it flies around a track.** *(Rolando Cerna)*

BELOW **A high-resolution orthophoto created by stitching together 644 photos taken on board a drone to create an aerial map of a farm. The images were processed with DroneMapper software. Each pixel of this image corresponds to about 4cm in the real world.** *(Author)*

Until recently, whenever you heard about drones it was usually in a military context or in some other destructive role. However, they have hundreds of other applications that can bring great good to the world, so in this section we'll cover some of their current and potential civilian applications.

Some people are constantly figuring out novel and useful ways to put drones to work for good, while others propose crazy applications for publicity purposes. Nevertheless, although the possibilities seem countless, they usually fall within a few main categories as discussed in this chapter. If you want to use your drone to make money, you'll first need to obtain the requisite licence from the Civil Aviation Authority (CAA), but for hobby and recreational use you don't need one. Of course, you'll still need to follow the regulations in terms of where you can fly. Further details regarding the safety aspects of flying your drone are described later on in this book.

Mapping and inspection

Probably the most important area for the commercial use of drones is for inspection and mapping work, where they're used to create high-resolution maps for use in surveying. Both fixed-wing and multirotor platforms are well suited to mapping applications. If you need to survey a larger region, a fixed-wing drone is the best option since it can fly further and faster. However, if you only need to survey a small area the convenience of being able to take off and land in a limited space might make a multirotor the better option. A multirotor is also better for some inspection tasks, as it can hover in place to take a good look.

Mapping, inspection and agriculture
Orthophoto

An orthophoto is an aerial map created from aerial photographs that have been stitched together with special software. An orthophoto

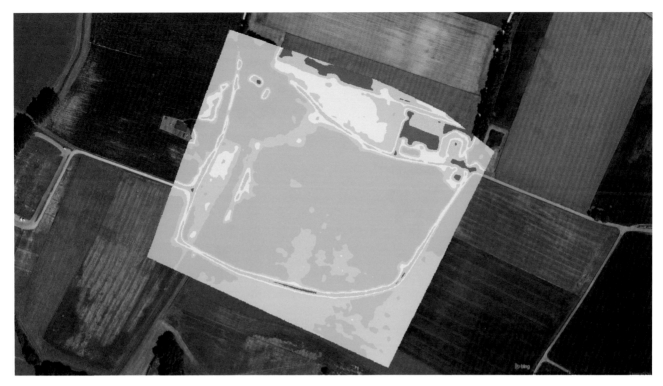

ABOVE An example of a smoothed NDVI image processed for a field – notice the roads don't have a high NDVI reflectance value (red) while most of the crops are relatively healthy (green), with the exception of certain areas. A farmer can use this type of image to determine what parts of their fields require special attention. The resolution of this image corresponds to each pixel representing 3.7cm in the real world, which provides very accurate information. The raw images are available on the Pix4D website example page, and the NDVI was processed by DroneMapper software. *(Author)*

produces a high-resolution image or map that's been geometrically corrected so that the scale is uniform. Often these images are defined by their spatial resolution, which is a measure of the area a pixel on the photo represents in the real world. Although such maps can be provided by satellite imagery, or by manned aircraft – which are able to cover much larger areas – the spatial resolutions of their pictures are often measured in terms of metres, with a single pixel representing at least one metre in real life. By contrast, orthophotos created from drones – which can fly much lower – produce much better sub-centimetre resolutions.

In order to get the best results when creating aerial maps, most autopilot systems have the ability to record the GPS location and altitude of each photo that's taken. This is often described as 'geo-referencing'.

Multispectral

Using cameras that are able to capture spectrums of light that we can't see – such as near-infrared – is particularly useful for agriculture. Near-infrared should not be confused with thermal cameras.

Most consumer cameras you can buy off the shelf can be modified by removing the IR filter from the sensor, which, though tedious, is a fairly easy task with most cameras. Modifying an off-the-shelf camera is often much cheaper than buying a dedicated near-infrared camera. The value of multispectral photography for agriculture is that it enables farmers to create an NDVI (normalised difference vegetation index) map of their crops, which can be used to check their health. (An NDVI image provides a representation of the amount of infrared light the crops are reflecting, with healthy plants reflecting the largest amounts of near-infrared light.) Creating NDVI maps require two cameras, one that takes regular photos and another that's been modified to take near-infrared images. By combining both and performing some basic calculations, the NDVI value can be found.

Digital elevation models

Digital elevation models are created by performing some fancy geometric calculations of matched points between several images to create a three-dimensional model of objects. This is useful for construction and mining applications, where a drone can fly over a given area and an accurate 3D model of the terrain can be created from the images captured. This 3D data can then be utilised to further analyse the area, such as by measuring the amount of earth moved in a mine etc.

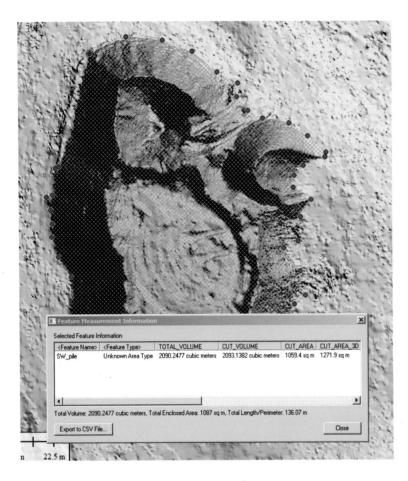

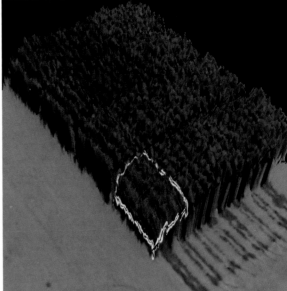

LEFT AND ABOVE Determining the stockpile capacity available for a mine from a digital elevation model created using DroneMapper software (above). Farmers can also estimate the growth of their crops by estimating the volume, provided the resolution is good enough. *(dronemapper.com)*

BELOW 3D digital surface created using images captured with a drone.
(Krzysztof Bosak)

Google Earth has recently been introducing similar technology to its maps, with many major cities now being rendered in 3D, as all their images are obtained from manned aircraft or satellites the resolution is much lower, which means the digital surface models are also less accurate (typically around 1m per pixel).

General inspection

There are some applications where all you need is a multicopter with a camera attached in order to visually inspect things that are normally difficult to reach, such as roofs, mobile-phone towers and other tall structures. One such application that's quite interesting is the use of drones to inspect offshore oil rigs.

Sensing

In addition to simply creating maps using cameras, you can attach special sensors to your drone to record all sorts of data as you're flying.

An example of a recent application of this sort occurred after the Fukushima nuclear

ABOVE The advanced airborne radiation monitoring (AARM) system integrates an unmanned aerial vehicle (UAV) with a lightweight gamma spectrometer and other positional sensors. *(Bristol University)*

power station disaster. In order to reduce the risk to humans in the aftermath, a multirotor equipped with radiation sensors was used to create an accurate radiation map of certain areas. Such tasks were previously carried out using helicopters flying thousands of feet above the ground, which produced radiation maps accurate to about 100m per pixel. By using a drone flying close to the ground, radiation maps accurate to within half a metre were produced. There are countless other sensing applications for drones, such as measuring mobile phone coverage in certain areas, or measuring air pollution. It's just a matter of integrating the appropriate sensors.

Aerial photography

Arguably the main application of drones is for aerial photography and filming, as they're able to produce high-quality footage at a fraction of the cost of using a manned helicopter. Using drones for photography is common in professional contexts, including many big budget movies, since in addition to the cost savings, it's also possible to fly drones much closer to objects and lower to the ground than a full-sized helicopter.

Aerial photography using R/C aircraft has been around since long before the existence of consumer drones, but the availability today of a degree of automation (thanks to GPS and flight controllers) means that aerial photography using drones is becoming ever simpler and more widespread. Many TV companies, including the BBC, frequently use footage captured by drones in many of their shows. Even in a hobby context the use of photography drones has become very popular, with many pilots capturing amazing footage in various holiday destinations.

Multirotors are the best-suited platform for aerial filming as they're able to lift heavier

ABOVE A hexacopter drone with a camera mounted on a gimbal allows for a dual operator set-up, where one pilot flies the drone while the second controls the camera and can examine areas independently of where the drone is flying.

cameras, and their ability to hover allows pilots to capture interesting angles and perspectives. There are numerous production companies offering aerial photography services using drones.

Surveillance

Surveillance is similar to aerial photography, but with more focus on monitoring events in real time. In addition to having an on-board camera, a real-time video feed is sent from the drone to the ground. Typical applications are for security purposes. Several public safety and emergency service units around the world have successfully used drones to help in emergencies and crises, while police use them to get an aerial overview of a situation long before a manned helicopter can arrive. Fire-fighters have used them to get closer to fires so that they can get a better idea of the situation without risking human life. Some rescue services have even used drones successfully to find lost or stranded people.

Other significant applications have been in conservation and anti-poaching roles, since a small fixed-wing drone with a live video feed

RIGHT Example of a custom-built fixed-wing drone used in the fight against poachers in Africa. *(Steve Roest)*

enables a single person to monitor large areas. Due to the remoteness of the areas in which they work, such usage generates fewer safety or privacy concerns regarding the use of drones for surveillance purposes.

Typically, most emergency service applications involve multirotor drones as these are able to take off and land in tight spaces, allowing them to get airborne faster. They're also often easier to transport than fixed-wing drones. However, for some security-based applications – such as monitoring an event or area – having a fixed-wing platform that's able to loiter for much longer and fly further is usually better, though these require more space to take off and land. Security drones usually have a camera mounted on a movable platform called a gimbal, which keeps the camera stable but allows the pilot to move it left/right or up/down to examine areas of interest.

ABOVE Specialist cameras can also be used, often referred to as block cameras, which are small, lightweight and allow control of the zoom from the ground. The one pictured here is capable of 36x zoom. *(Author)*

Delivery by drone

The possibility of using drones to deliver parcels has recently received a fair amount of media coverage after Amazon declared its intention to use them for small parcels. Although most of the technology is available to achieve this, there are still some important hurdles to tackle before it could become a reality. Current battery technology means that most multirotor drones can't fly long distances, which limits their practicality in a delivery role. There are also numerous safety issues when operating drones over built-up areas potentially crowded with people, not to mention other air traffic. Also, the over-reliance of all drone systems on GPS technology is an important issue – for instance, if the GPS signal were to fail or be blocked all drones would essentially have to fly blind, as they'd have no way to estimate their position. Systems are being designed to overcome this, but they're still largely military and unavailable to the public.

It can be argued that the real benefit of drones for deliveries will be in the improvement

LEFT Example of a multirotor drone modified to carry parcels for delivery. *(Frank Höffner)*

of services to remote places. Some companies, including Google, are working on using them to deliver important items such as medicine to remote areas and places hard to access using ground vehicles. One novel suggestion is that an 'ambulance drone' network could be established in which drones could be despatched to any location within minutes to provide important first aid supplies before a paramedic could get to the scene.

So who knows that the future may hold? Maybe the next time you order a drone it will fly itself to your door direct from the factory!

Hobbyists

Although there are many ways to use drones for good, there's also lots of room to use them just for fun! The radio-control hobby has been around for a long time, but only in recent years has its adoption of quadcopters and drones become more widespread. The use of sensors and flight controllers has opened the hobby up to many new pilots, as it makes flying easier and safer. With some drone platforms being technically able to fly themselves, just about anyone can fly one. Drones are also popular amongst model-makers, who find the idea of programming and tinkering with a flying electronic device enthralling. So the drone hobby has a collection of both traditional radio-control enthusiasts and electronic geeks! Depending on where you fit into this spectrum, the drone hobby has low entry requirements – thanks to drones being super-easy to fly – as well as a limitless list of possibilities, whether you want to get your hands dirty and delve deep into the code or design your own custom aircraft.

Aerial photography

As previously discussed, advances in technology have opened up the world of aerial photography to anyone using a drone, so that even hobbyists can now capture the world around them from above. Some autopilot systems also have the ability to automatically follow you, or circle around you while recording whatever you're doing, thereby allowing you to record yourself hands-free. A word of warning though – always make sure the altitude is set high enough that it doesn't fly into any trees!

The term 'dronie' is also gaining popularity, which is basically a selfie photo taken from the unique perspective of a drone, often captured by the video of the quadcopter as it flies away and upwards from the subject. However, please always make sure you follow the rules and regulations set out in Chapter 6 when trying to capture something unique with your drone. The general rules are not to fly close to people or over built-up areas.

First-person view (FPV)

First-person view or video piloting is, as the name suggests, flying your drone as if you're sitting inside it. Using a camera mounted on your drone to send live video back to the ground allows you to get a total sense of

RIGHT On-board view of an FPV aircraft. The camera is on a pan-tilt mount that allows the pilot to move it around as if they're actually sitting inside the aircraft.
(James Whomsley)

LEFT FPV gear commonly seen at the flying field, including FPV goggles that the pilot wears to view the live video feed from the aircraft. A portable chair is also useful as it makes the whole experience more comfortable.
(David Fant)

immersion while flying. More advanced FPV set-ups can also include a pan-tilt camera, which is linked to the pilot's video goggles so that the camera motion is synced to wherever the pilot looks – so if the pilot looks down the camera aboard the drone will do the same.

FPV drone racing

Drone/mini quadcopter racing is becoming ever more popular. This involves racing small quadcopters – typically less than 25–30cm in diameter – through a course, piloted via the first-person-view video feed. The agility of these low-cost mini racing quadcopters means you can fly them through small spaces and perform acrobatic stunts with them. Many courses involve zigzagging between trees, flying under hoops and negotiating other obstacles. Drone racing leagues are also forming, with several FPV racing events happening around the country and world.

Some might argue that mini racing quadcopters aren't true drones, since they can't fly entirely on their own. This is because most of them use a basic flight controller that only keeps the copter stable, while it's remote-controlled by a pilot. I won't get into this argument, as many people call them racing drones while others refer to them as mini quadcopters. You can choose for yourself what you want to call them!

Research and teaching

Let's not forget that drones can be excellent platforms for teaching and research tasks. Many drone systems are completely open source, allowing you access to the source code and design documentation. This means that you can learn how things work in a very practical manner.

Drones cover three core areas of engineering: mechanical, electronic and software. Mechanical expertise is needed when designing and constructing the airframe; electrical engineering skills are important when using the actual autopilot boards and other electronic components; and software engineering comes into play when working on the code that controls everything.

Many universities and enthusiasts around the world use open source drones on specialist projects and for the general development of drone technology. Because all this research is shared through the open source licence, it means that you don't need to reinvent the wheel each time you start work on a new project.

BELOW Inspecting the source code of the Flip32+ flight controller board to find out exactly how it works.
(Author)

CHAPTER 4
ANATOMY OF A DRONE

(Author)

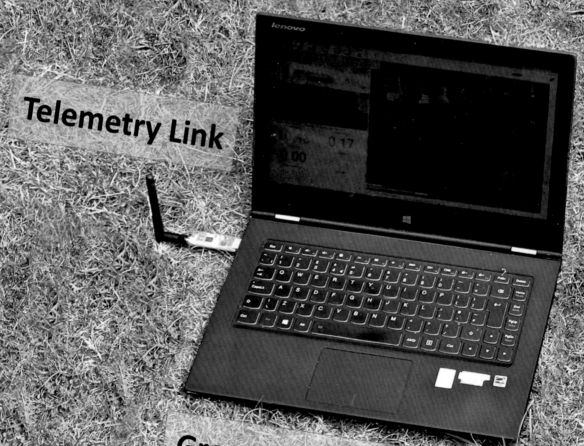

Telemetry Link

Ground Control Laptop

OPENING SPREAD
A standard hobby drone set-up consists of the drone itself – multirotor or fixed-wing – a ground-control laptop, an R/C controller and a video monitor. On board the drone are the autopilot, camera, battery, telemetry link, video transmitter and other equipment. The R/C controller is used to manually control the drone; the video monitor receives real-time video from on board the quadcopter so you can see what you're recording; and the ground-control laptop is used to send commands to your drone, such as setting waypoints. The laptop communicates with your drone via the telemetry link.

A standard drone set-up starts with the drone itself, which could be either a multirotor or a fixed-wing plane. The drone platform includes an autopilot, camera, battery, telemetry link and video transmitter. On the ground you'll have your R/C controller, which you'll use to manually control the drone, while other equipment can include a video monitor that lets you receive a live video feed from the drone's camera. Drones also use a ground-control laptop (or tablet) that you can use to monitor its systems, such as altitude and speed, and can include other features such as a map that shows you where your drone is flying. You'll also use this ground station to set waypoints and send commands to your drone, such as telling it to take a photo. Connected to your ground station is a telemetry link that wirelessly sends and receives all the information between it and the drone.

Airframe

Fixed-wing airframe

We've already discussed the main types of aircraft that are used for drone platforms, but in this section we'll go into more detail about the actual materials used, and other factors you need to consider when choosing and building a fixed-wing aircraft. We won't cover the details of building a fixed-wing aircraft from scratch, as that alone would take an entire book – if you're keen to do this you should pick up a general radio-control model aircraft book, which will tell you everything you need to know. It's something that I'd suggest you should certainly try at a later stage, as it's very rewarding to see something fly that you've designed and built from scratch.

Airframe materials

When it comes to material selection for a fixed-wing drone the main ones include foam, balsa wood and composites such as fibreglass and carbon fibre. However, almost all hobby fixed-wing drones are made from foam-based materials, as they're cheap, lightweight and easy to repair with some glue.

Other materials could include foam with a plastic covering to improve the robustness of your aircraft, but many people simply add packing tape to the wings and belly to protect the foam during landing. Although this will add some extra weight it increases the lifespan of your aircraft, particularly when you land on rough surfaces such as gravel.

Some of the higher-end platforms may be made from carbon fibre or fibreglass, as these provide the greatest strength and durability. However, I'd suggest you only move on to these materials once you have more experience with flying and building drones, as they're much harder to work with, and if you have a crash it's unlikely that you'll be able to repair your aircraft quickly or even at all, as it's not as simple as just gluing two pieces of foam together.

Payload area

When it comes to choosing your fixed-wing drone airframe the payload area is an important factor, as this is where you'll mount all of your equipment. Firstly you must make sure it'll be big enough to hold at the very least your autopilot, battery and camera. More importantly you must also ensure that the aircraft has enough space for your equipment to be positioned in forward or rearward locations, to ensure the aircraft's centre of gravity is correct. With most aircraft the centre of gravity will be about one-third of the way back from the leading edge of the wing. This is usually marked on the aircraft or mentioned in the assembly manual.

It's difficult to calculate the exact centre of gravity value for a wing without going into technical details of the aerodynamics, but the 'one-third of the way back' rule of thumb is a good enough estimate for most aircraft configurations. Manufacturers often calculate this value and provide a recommendation for you. There are several tools available on the Internet that can help with your centre of gravity calculations, but in general you should always make sure the centre of gravity is in front of the centre of lift of the wing, so that your aircraft is slightly nose-heavy. The reason for this is that if it stalls it will tilt downwards, allowing it to gain speed and recover.

Servos

Servos are primarily used on fixed-wing aircraft as actuators to move the control surfaces, but they can also be used for other tasks that require some form of actuation, such as tilting

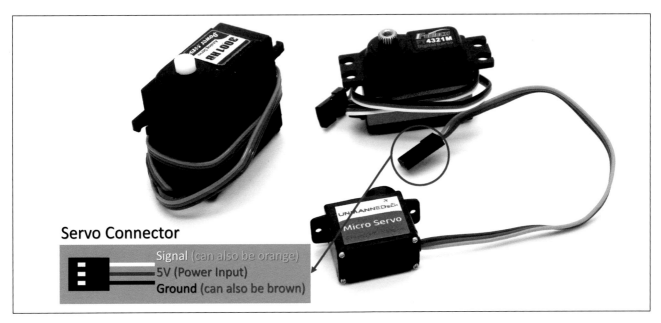

a camera up and down. On fixed-wing aircraft each control surface typically has a single servo to control it, so you'll have a servo for the rudder, a servo for the elevator and so on.

The gears inside servos are either plastic or metal. Plastic-gear servos are cheap and lightweight, but if you require more demanding tasks of your drone metal-gear servos are better, as they can produce more torque than plastic. You must ensure that the servo you're using is powerful enough for its given task, or else you might break the servo gears or burn out the motor inside it.

Detachable wings

Depending on the size of your aircraft you might want to look for one that has detachable wings. These make life much easier when transporting and storing a drone, as some wings are quite awkward to fit into a car if you're unable to remove them.

Multirotor frames

The great thing about most multirotors, particularly quadcopters, is their mechanical simplicity, which means that they're very easy to build using basic materials you can easily find at your local DIY shop, such as wood. However, scratch-build quadcopters are becoming less common due to the ever-reducing cost of multirotor frames. Starting with a cheap purchased frame is often a better idea for your first quadcopter, as it means you don't have to worry about making something that turns out to be too heavy, or too weak, and allows you to focus on actually building and learning rather than trying to design a suitable frame.

Building your own frame could nevertheless be a fun process to try once you've learned more, and thanks to the emergence of affordable and easy-to-use 3D printers and CNC machines, making a scratch-build quadcopter out of plastic parts and carbon fibre is much easier now than it was just five years ago. If you have the time to learn, designing and building your own multirotor can be a rewarding experience.

Central plates

The central plate is the main section of a multirotor on which the arms are mounted, and

ABOVE Examples of servos commonly found in the R/C hobby. *(Author)*

BELOW Most of your equipment will be mounted in the central plate area of your multirotor drone. *(Author)*

RIGHT A selection of the various types of arms you can use on a multirotor. (Author)

it's often constructed out of fibreglass or carbon fibre. When choosing your platform it's important to make sure the central plate has enough room to mount all of the equipment you'll need, such as the autopilot, R/C receiver, telemetry equipment and so on. Some multirotor frame designs have multiple central plates one on top of another to create more room for all your gear, often referred to as stack-up plates.

Arms

The arms of a multirotor are what the motors are attached to. Depending on the frame, arms can be made from plastic, fibreglass or carbon fibre. Most take the form of either square-section or circular tubes, but on some smaller multirotor frames they can be flat plates (particularly common on mini quadcopters) or other plastic structures. Some multirotors can have their arms folded away to make it easier to store and transport them.

It's often nice to have the strongest arms possible on your multirotor, but this will increase the frame's weight significantly, which will in turn negatively impact on your flight times. Consequently when buying a frame the manufacturer will often tell you the maximum take-off weight that it can comfortably support, which will give you an idea of the arm strength and what you can carry on your drone.

Power distribution board

Since multirotors have more than one motor and electronic speed controller (ESC), connecting power to them can get quite messy and complex if you need to solder all the cables together. For this reason many frames include a power distribution board (PDB) built into the central plate, making it much easier to power all your equipment and motors. Some PDBs have extra features that might include a voltage regulator, which will convert your battery voltage to 12V or 5V so that you can power other equipment on your drone, such as FPV gear, gimbals and cameras. Others also include connectors for your receiver and autopilot to make the wiring much neater. If your frame doesn't include a power distribution board you can always purchase one separately.

When purchasing a PDB make sure you

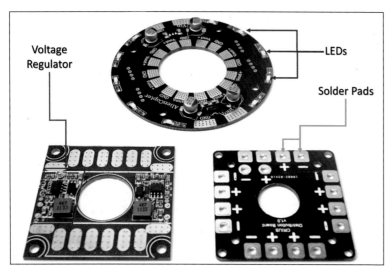

BELOW Examples of power distribution boards used on a multirotor drone to route power between the various ESCs and motors. Some PDBs also include a voltage regulator to output constant voltage that you can use to power auxiliary gear such as your FPV equipment or LEDs. (Author)

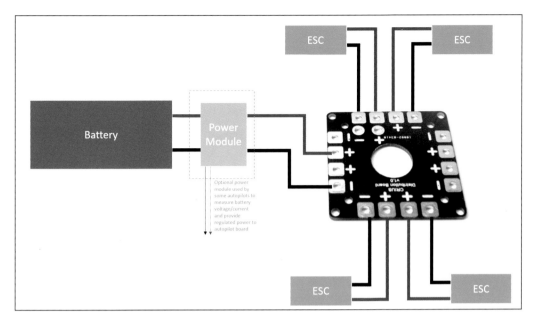

LEFT A PDB allows battery power to be connected to several other devices, such as your motor electronic speed controllers (ESCs). *(Author)*

check its maximum current rating, and compare that to the maximum current draw of all your motors added together to make sure it's suitable. Alternatively, as a very conservative rule of thumb just add up all the ESCs you're using on your multirotor; so if you're using four 20A ESCs on your quadcopter you'll need a PDB rated to about 80A. In reality, however, the total current flowing through the PDB is much lower than this, as your motors and ESC won't be working at full capacity the entire time. Consequently it's always best to use the actual motor current draw values included in the motor's specification.

Payload area

If you plan on carrying some equipment, such as a camera and gimbal, you'll need space to mount it on your multicopter. Some frames are designed to have the batteries mounted underneath the central plate, so mounting a gimbal in addition to the batteries on such frames could be tricky. Others are designed to have the battery mounted on top of the central plates. This keeps the bottom free, allowing you to easily mount your gimbals or other gear underneath the multirotor frame. Lastly you'll also see some frames that make use of load booms as a method of mounting payloads. Load booms comprise two fitted horizontal pipes.

Monocoque

This is a fancy name for a very lightweight and sturdy frame that doesn't feature any internal skeleton, as the exterior skin itself forms its structure. This type of frame is seen on many plastic ready-to-fly drone platforms, as with all its insides hidden it makes for a neater and more professional-looking drone. It can also provide extra protection from the elements. However, when it comes to a DIY drone platform that gets tinkered with, having to remove the entire top half each time you need to check or change something can be a bit annoying!

Landing gear

Landing gear are like the legs of your aircraft. On multirotors they're often used to prevent equipment mounted underneath – such as cameras – from hitting the ground during landing. Most are fixed, but you can also fit

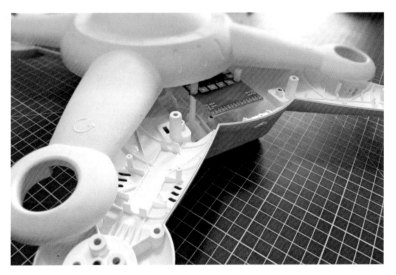

BELOW A monocoque quadcopter frame comprises a shell-like body consisting of top and bottom halves. This shell provides the structural rigidity and casing for the drone. Components such as motors and ESCs are mounted directly to the bottom half of the frame. *(Author)*

ABOVE Since most hobby drones are hand-launched by throwing them into the air, having landing gear isn't necessary as it just adds extra weight and complexity. Most of the time landings are on grass, which won't usually damage the aircraft. *(James Whomsley)*

retractable landing gear that can move up and down. Most fixed-wing drones don't use landing gear at all, but land directly on the fuselage, which skids across the ground during landing. Landing gear are only really used on heavier fixed-wing drone aircraft, as most of the time you'll be landing on grass. However, if you often have to land on more rough surfaces it's worth adding reinforcement to the belly of your aircraft to provide some protection.

Retractable landing gear

Retractable landing gear are mainly useful on fixed-wing aircraft, as they can be tucked away inside the wing during flight to reduce drag. On multicopters the main purpose of retractable landing gear is to fold them out of view of the camera so that you don't see them on the edges of your photos or videos. It also enables the camera to rotate without the legs getting into the picture. However, retractable landing gear is only recommended if you need it for filming purposes, as it's heavier and more costly than equivalent fixed landing gear.

Ground clearance

Another aspect to consider is the ground clearance your landing gear will provide, especially if you plan to mount equipment such as a camera under your drone. If you do you'll need to ensure there's enough ground clearance to allow for it.

Ground contact area

Depending on the types of surface on which you'll be landing, the size of your landing gear's contact area with the ground will be affected. For example, when landing on softer surfaces – with snow being the most extreme example – some landing gear won't have enough ground surface area to stop a drone from sinking right in. Because of this many pilots add some foam tubing to their landing gear to increase its contact area.

Autopilot

The autopilot can be considered as the brains of your drone, as it processes all the information and sends the relevant commands to the motors and control surfaces to perform

RIGHT Having landing gear that can retract (as on this aircraft) means that you can look down with the camera without the legs showing up in your video. *(Author)*

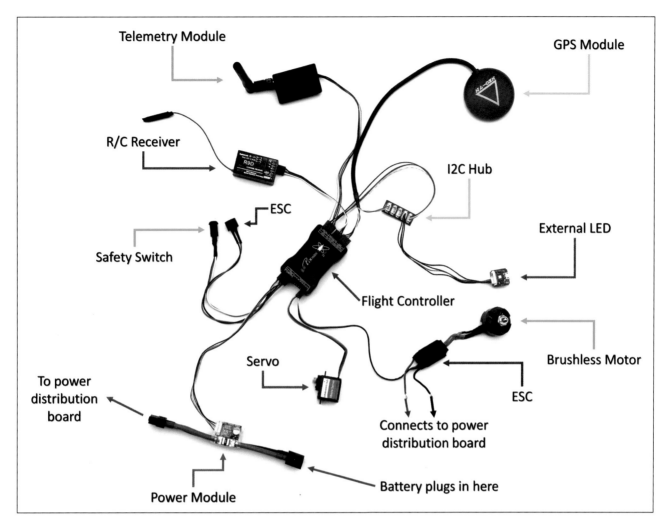

ABOVE A fully featured autopilot system consists of the main flight controller and several other pieces of equipment connected to it. Multirotor drones have several ESCs connected to the flight controller, whereas a fixed-wing aircraft would have more servos connected to it. *(Author)*

the desired actions. Most autopilots include sensors such as gyroscopes, accelerometers, barometer and a magnetometer, which are used to measure the aircraft's motion as it's flying. The autopilot will use this information to keep the aircraft flying and heading towards the target you've set for it by controlling the motors and other control surfaces. In this section we'll discuss the autopilot in some detail so that you have a better idea of how it works.

Flight controller

When we refer to the autopilot system we include all of the main components, including flight controller, external GPS sensors and telemetry. The flight controller is the main component of the system and does all of the calculations to keep your aircraft flying, but it doesn't necessarily have the ability to control fully autonomous flight – sometimes additional sensors such as GPS are needed for this. There's a huge selection of flight controllers available to buy and the choice can be overwhelming at first. To help inform your decision when selecting one this chapter will discuss some of a flight controller's key features and how they work.

It's worth noting that some flight controllers only have the essential sensors, such as an accelerometer and gyroscope, as extra sensors aren't needed and just add to the cost of the unit. For platforms such as racing quadcopters the flight controller is just used to help keep the aircraft stable while the pilot controls what it's doing, so additional sensors like a barometer or GPS are often not used. These basic flight controllers are great as they're much cheaper and easier to set up, since their main purpose is to stabilise your aircraft, not enable it to perform fully autonomous operations. Some flight controllers even include a built-in LCD display and buttons allowing you to set up

ABOVE The circuit board of an autopilot. *(Author)*

and configure your aircraft without needing to connect it to your PC, as it's all done on-screen.

More advanced flight controllers support more sensors, like GPS, and other accessories that enable much more advanced functions, such as the ability to automatically take off and land your aircraft or fly to specific waypoints and perform actions automatically.

Accelerometers and gyroscopes

These are the most important sensors of your flight control unit as they measure the aircraft's angle and rotation and keep it stable. An accelerometer is essentially a small device that measures gravity. Most modern-day accelerometer chips actually include three accelerometers built in to measure all three axes (X, Y, Z). By coupling together the measurements of all three accelerometers the current angle of the aircraft can be measured. For example, if the aircraft is level the Z axis will measure the same force as gravity and the X and Y axes will be zero. Similarly, gyroscopes measure acceleration or rotation, and just like accelerometers most gyroscope chips can measure all three axes. Some manufacturers have even managed to combine a three-axis accelerometer and a three-axis gyroscope into a single chip. These accelerometer and gyroscope chips are often the same as what you'd find inside your smartphone.

Unfortunately in real-world applications sensors such as gyroscopes drift over time, and since accelerometer readings are distorted when your aircraft is turning (due to centrifugal forces) it can be difficult to keep track of the aircraft's true angle. Without going into any confusing maths, the autopilot uses some fancy tricks to estimate errors and correct for them. Some also combine information from additional sensors like GPS, barometer and compass to get more accurate readings. This process is known as sensor fusion. However, the barest minimum of sensors required to estimate an aircraft's angle to an acceptable level while it's flying are accelerometers and gyroscopes.

When you're first setting up your flight controller you'll need to calibrate the accelerometers, since the processes of manufacture and transport could leave the alignment of the sensors slightly 'off'. This means that when your autopilot is positioned on a level surface it might measure itself as being slightly tilted. The process of calibration will correct for these slight misalignment issues via software, and is a very common requirement. To get the best results during calibration you should use a bubble level to ensure the autopilot is exactly level when you calibrate the sensors. After you've performed the accelerometer calibration successfully you won't need to do this again unless you update the firmware on your flight controller.

When you turn on your drone before flying it's very important that it isn't moving at all, so that the accelerometers and gyroscope sensors can initialise to get the best performance. Sometimes if you turn on the drone while on a moving platform, such as a small boat, the initialisation can be slightly off. You might still be able to fly it, as your autopilot uses other sensor information such as GPS to keep it stable, but for best results it's always best to make sure your aircraft is perfectly stable during the initial start-up.

Barometer

The barometer is a very sensitive pressure sensor on the autopilot that's used to measure your aircraft's altitude. As the aircraft ascends or descends the air pressure changes, and the barometer measures this change, which corresponds to a specific change in altitude. Most barometer sensors include a built-in temperature sensor to compensate for any change in pressure due to temperature, and most barometers used on autopilots today are able to measure sub-centimetre changes in altitude readings.

Although changes in pressure provide a great indication of changes in altitude, they aren't always very accurate in measuring the absolute altitude of your aircraft relative to sea level. For this reason the altitude information from your GPS sensor is combined to get a better estimate; then, when you take off, the autopilot will use the altitude above sea level recorded from the GPS as a calibration point and measure the slight changes relative to that point with the barometer.

Magnetometer (compass)

The compass sensor measures the magnetic field around your aircraft and allows your flight controller to know what direction the aircraft is facing, just like a regular compass. On a fixed-wing aircraft it's very easy to tell the direction of travel since it can only fly forwards, so compass sensors are only essential for multirotor platforms, as they're able to hover.

As you can imagine, having a compass on board an aircraft that has motors with big magnets and other electrical wires means there's a lot of magnetic interference on board, which will cause problems with the compass sensor. For this reason the compass is often located on the actual GPS module and mounted far away from all the other electronics to minimise interference.

Because the magnetic field is different depending on where you're flying, it's important to perform a quick compass calibration when flying in a new area to get the best performance out of your drone. This is often referred to as a 'compass calibration dance', as it involves rotating the multirotor about each axis, which might make you look slightly strange to an onlooker. During compass calibration make sure that you stay far away from any metal objects or power lines.

A common indication of poor compass calibration or too much interference is demonstrated by what's known as the 'toilet bowl effect', aptly named after the way water spirals around a toilet when it's flushed. This effect is noticeable when you enter a 'position hold mode' on your multirotor, which, if your compass is getting too much interference and your flight controller can't keep track of its true orientation, will cause it to spiral around the 'position hold' point in a gradually bigger and faster circle. Should this happen during a test flight simply switch back to manual mode, fly the multirotor back and land.

ABOVE GPS/compass modules on multirotors are usually mounted on a mast to keep them away from extra interference. *(Author)*

GPS (Global Positioning System)

GPS receivers are the devices that provide your drone with positioning information so that it knows where it is. GPS works by measuring the time it takes for a signal sent from a GPS satellite to arrive back at the receiver. By measuring the distances from several satellites a 3D position can be found. However, these signals are distorted and sometime bounce off intervening objects, which means that they take longer to arrive; so GPS isn't perfectly accurate, measurements typically being accurate only to about 5m–10m horizontally and about 15m vertically. The most useful thing about GPS is that it provides a great absolute 3D position of your vehicle relative to the Earth, and in common with other sensors such as the barometer, which measures relative altitude, this means that by combining data from all of them the position of your drone can be estimated fairly accurately. The more satellites you're connected to will result in better position estimations.

Before you take off, as the autopilot is

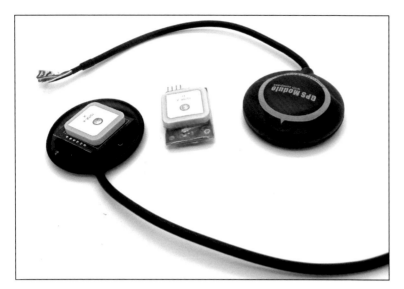

ABOVE GPS modules come in a variety of shapes, sizes and modes. Typically the larger modules get a better signal, due to their larger antennae.
(Author)

starting up it'll wait to connect to sufficient GPS satellites for the initial position to be recorded and will save this take-off location as its home point. So if something goes wrong or you want the drone to return home it will land at that same point or within a few metres (due to GPS discrepancies, as explained above).

To further improve the accuracy of the GPS position many of the more modern GPS modules are able to connect to more networks, such as the Russian GLONASS network, which greatly increases the number of possible satellites to which the module can connect, which in turn increases accuracy and reliability. So it's always better to use a GPS module that can support both American GPS and Russian GLONASS networks for best performance.

When mounting your GPS module it's important to make sure it has a clear view of the sky so that it can receive signals from the GPS satellites. Often on multirotors you'll find the GPS and compass module mounted on a small mast to ensure they're clear of any interference. On fixed-wing aircraft it's also important to make sure it's mounted with a clear view of the sky. If your aircraft is made from foam then you can mount the module internally, as GPS signals can easily travel through foam. But if you have a fibreglass or carbon fibre aircraft you might need to mount it externally, as GPS signals don't pass through these materials very easily.

Power module (current and voltage sensor)

Most autopilot systems include a power module, which is used to provide a filtered and regulated power supply to your flight controller directly from the flight battery. Most electronics run at 3.3V or 5V, so the power module will convert your battery voltage (which is often much higher) down to the 5V that's fed into the flight controller. The power module will often include a current and voltage sensor that measures how much battery capacity your drone has left, much like a fuel gauge in a car. Some flight controllers are also able to estimate the remaining flight time and are smart enough to automatically take over, return your quadcopter home and land it before your battery runs out.

However, with most build-it-yourself drones it's important to properly set up your current and voltage sensor for the battery that you're using. This is often a relatively simple process that's documented in the flight controller manual, and usually just requires you to specify the cell count (voltage) and the mAh (milli-Ampere hours) capacity of your battery.

Also, before connecting the battery to your power module make sure that the battery voltage isn't higher than the module's supported voltage, otherwise you can damage your module or autopilot. You'll also need to make sure that you connect the battery to the correct input, as power modules often have input and output connectors.

Distance sensors

The addition of distance sensors provides added functionality and safety features to your aircraft. Most distance sensors used on drones are ultrasonic. These measure the time it takes an ultrasonic ping to bounce back from an object to determine its distance. Other more expensive and more accurate systems use lasers to measure the time it takes for a beam of light to bounce back off an object. These laser-based systems also have a significantly greater range than ultrasonic-based sensors. The most common application of such sensors is to measure an aircraft's distance above the ground, which is particularly useful when coming in to land or during take-off. However, they can also be mounted to look ahead of the aircraft to avoid it crashing into obstacles.

Ultrasonic sensors work best when flying over hard surfaces such as concrete, as more natural materials such as grass can deflect and absorb the sound pulses, thereby reducing their accuracy.

Optic flow sensor

Almost all drones rely on GPS satellites for their positioning information. However, this means that when flying indoors, close to tall buildings or under large trees the GPS signal is often too poor for reliable positional information. Consequently some autopilot systems support the use of optic flow sensors to help maintain the aircraft's position automatically, enabling your multirotor to autonomously hold its position without GPS. An optic flow sensor is essentially the same type of sensor as used on a computer mouse, which is a low resolution camera that records the motion of pixels across it to estimate its motion. However, optic flow sensors only work at low altitudes (below about 10m above the ground) and when the surface has sufficient contrast, such as grass.

Airspeed sensor

In order for a fixed-wing drone to estimate its speed it can rely on the GPS module, but this only provides the ground speed (the distance travelled along the ground between two points in time). But since there's often wind the speed of air flowing over the wings of an aircraft doesn't always match the ground speed. This is referred to as airspeed. There's a slight difference between indicated airspeed (airspeed measured from a sensor) and true airspeed (dependent on atmospheric conditions and altitude), but we don't need to go into detail about that.

If you're flying on a calm day the airspeed and GPS speed will be very similar, so some fixed-wing autopilots don't have an airspeed sensor. However, if your drone is flying directly into a very strong wind the ground speed can be very low but the airspeed will be much higher. Having an airspeed sensor therefore provides the autopilot with the actual airspeed of your drone, which is important in preventing a stall.

An airspeed sensor is a differential pressure sensor that uses a pitot tube sticking out of the front of your drone. By comparing the static and dynamic pressure values from the pitot tube the airspeed can be calculated.

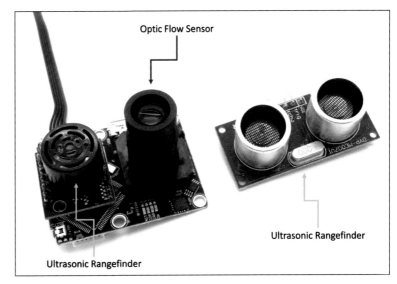

ABOVE Ultrasonic sensors use high-frequency sound pulses to measure the distance to objects, whereas an optic flow sensor uses a camera to measure its motion over a surface. *(Author)*

You never need to use an airspeed sensor on a multirotor since it doesn't stall, as the motors provide all the lift to keep it flying.

Flight modes

Flight controllers often have the ability to set various flight modes. This is useful, as it allows you to change the behaviour of your drone while it's flying. The flight modes are normally controlled through your R/C controller via one of the switches. The actual modes will vary depending on what flight controller you're using, but these are the basic ones:

Stabilise mode – The most basic flight mode and often referred to as 'manual mode', this will keep your drone level and stable. You're still able to manually fly the drone,

BELOW An example of an airspeed sensor that can measure how fast your aircraft is travelling through the air. *(Author)*

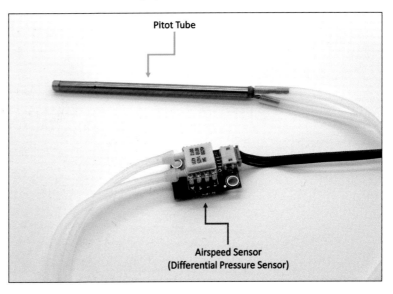

but if you let go of the controls the drone will automatically self-level. For safety reasons you should always have this set as a flight mode just in case something goes wrong, such as you set a waypoint that's directly in the path of a tree; then you're still able to resume manual control.

Acrobatic mode – This is designed for acrobatics, and makes the response of your drone much more sensitive. Often acro mode doesn't self-level your drone, but rather locks its current orientation so you can perform various acrobatic manoeuvres.

Altitude hold – This is like stabilise mode, but will also control the altitude of your drone to keep it flying at a specific height.

Position hold – This flight mode uses the GPS sensor to keep your drone flying at a set location. Fixed-wing aircraft will circle this point, but multirotors will hover at the specific position.

Return to launch – When this flight mode is triggered your drone will fly to a set altitude and return to the take-off position. Some flight controllers will also control your drone to make it land at its starting location. This flight mode is useful in the event of an emergency or if you can no longer see your drone. Often you're able to set it as a failsafe on your flight controller so that if the signal with your R/C gear is lost it will automatically fly home.

Autonomous mode – In this mode the autopilot will control your drone and fly it for you. Usually this involves setting a mission with waypoints for the drone to follow, which can also include automatic take-off and landing. Some autopilots also let you perform specific actions at waypoints, such as retracting the landing gear or taking a photo.

Data logging

Some flight controllers have the ability to record data from sensors and what the flight controller is doing, so that you can review them after a flight – exactly what the black box (that's actually orange!) does on manned aircraft. Many flight controllers will save the flight data logs on to a secure digital (SD) card, while others might have some internal memory that will store this data. Recording the flight data is a great feature if you ever have a crash, or something isn't working as it should, as it allows you to look in detail at what was happening so that you can find the cause of the problem or the reason for the crash. Some autopilot systems, such as the ArduPilot platform, also have PC software that will automatically analyse your flight logs to detect common problems.

BELOW Inspecting some of the data stored on an autopilot log file allows you to see exactly what's going on inside the autopilot if something isn't working correctly.
(Author)

Vibrations

Just as with cameras, vibrations can cause problems with your flight controller. Irrespective of what platform you're using it's always a good idea to mount your flight controller on an anti-vibration mount to ensure that sensors such as accelerometers and gyroscopes don't get corrupted readings due to vibrations.

PID tuning

Just about every electronic device that controls a physical object will most likely use a control loop feedback mechanism called a PID controller, which stands for 'proportional integral derivative'. To explain what this does we'll take the example of a quadcopter that's trying to keep level. If a sudden gust of wind causes the quadcopter to roll to one side, the PID controller measures the difference between the quadcopter's current roll angle and its desired roll angle of zero. Based on this error the PID controller will determine the best commands to send the motors to get the quadcopter level again as fast as possible without overshooting or oscillating about the level flying position.

PID controllers are used in many areas of the flight controller, including keeping a fixed altitude and staying in a particular GPS position. PIDs are also used in ESCs, servos, gimbals and many other pieces of equipment. Often the default PID values set on a given flight controller are adequate for your needs, so you might not need to worry about PID tuning, but as you become more familiar with your flight controller you may prefer to tune the PID values to get the best possible performance out of your drone.

PID tuning can be quite difficult to achieve, but you can use the information in this section to help you get started.

Proportional term

The 'proportional term' will send a correction that's proportional to the amount of error. Going back to our example of a rolling quadcopter, if it's further away from its level position it will tell the motors to work harder to get back to level flight. Having a P-term that's too low will result in an unstable aircraft, as the correction will never be strong enough to bring the aircraft back to the level position. Increasing the value of P will affect how strong the correction will be, hence speeding up the response of your quadcopter; however, if the P value is too large it will overshoot the level position and end up being tilted in the other direction. To reduce the amount of overshoot we can reduce the P value, but this could result in the quadcopter taking too long to return to the level position; so the 'integral term' is introduced.

ABOVE Using an anti-vibration mount can greatly improve the performance of your autopilot, as the inertial sensor readings are then less distorted. *(Author)*

BELOW An example of the effect of varying PID values on the time it takes to reach a stable point. The blue line represents the desired reference position. The purple line represents the response if the P value is too high, which causes oscillation around the reference. The red line is when the values are too low, and although there's no oscillation it takes too long to reach the reference. The green line represents the ideal response, with perfectly tuned PID values. *(Wikimedia Commons)*

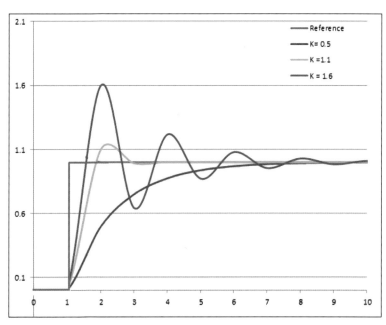

Integral term

The 'integral term' or 'I-term' essentially controls the maximum correction amount that can reduce the amount of overshoot; this is based on how much error there is between the current position and the desired position. The integral term will increase or reduce the correction depending on how far away from the reference it is and how quickly it's changing. You can think of the integral term as a way to dynamically vary the P-term depending on the situation. So in the case of our quadcopter that's been disrupted by a gust of wind, as the error is initially very large the integral term will ensure a large correction signal is sent to the motors to start returning the quadcopter to level flight as soon as possible, but as it approaches the level position it will reduce the control signal to avoid overshoot.

Derivative term

The 'derivative term' controls the speed at which the aircraft returns to its reference position. The D-term isn't often used with many PID controllers, as setting P and I values will result in adequate performance. Setting a higher value for the derivative term will cause your aircraft to be more sluggish, as motions will be slower.

Best PID tuning process

In order to tune your PID controller, this is the recommended process. You should start by gradually increasing the P-term until you notice some oscillation. Once you reach that point, reduce the proportional term slightly. Next tune the integral term, and again increase it slowly until you notice some oscillations, when once again you should reduce it slightly until the oscillations stop. If your system has a zero value for the derivative term by default you'll not need to worry about it. However, if it does have some derivative value set, to tune it you'll need to reduce the D value slowly until you notice your vehicle becoming unstable, then increase it slightly to the lowest point you can before it becomes unstable again.

To better understand how PID values change a system there are a few interactive PID simulators on the Internet that let you play with the values to see how they react.

Automatic PID tuning

Many recent flight controllers include a method to automatically calculate the optimum PID values for your vehicle by performing an automated test flight. However, if you do this with your aircraft, then in order to get the best results you'll only be able to do it on a very calm day with no wind. You'll also need to make sure your aircraft is in a flyable condition. Usually your flight controller's default PID values are adequate.

Ground station

Having a ground station isn't essential for your drone, but it does complete the entire system, allowing you to monitor and change your aircraft's parameters as it's flying or simply to view a live video feed. A ground station consists of all the equipment you'd use on the ground for your unmanned aircraft.

BELOW Due to its easy portability an android tablet is often used as a ground station to communicate with drones. *(Creative Commons License)*

RIGHT **A telemetry system consists of an air module, which is on your drone and connects directly to your autopilot, and a ground module that connects to your laptop/tablet.** *(Author)*

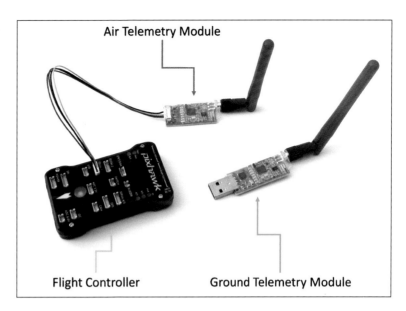

Telemetry

Telemetry modules are small radio devices that communicate between each other, allowing you to talk with your autopilot as it's flying. Most telemetry systems allow communication in both directions, enabling you not only to receive information from your autopilot but also to send it commands remotely from the ground. A basic telemetry system consists of two modules, one located on the aircraft and the other on the ground. They often transmit on the 433MHz frequency band, which since it isn't used by your R/C gear (2.4GHz) or your FPV gear (5.8GHz) minimises the possibility of interference.

433MHz telemetry modules often have a range of over 1km, making them the ideal choice for most systems. However, there are other telemetry solutions – based on Bluetooth – that are only useful for short-range connection, mainly to change the parameters on your aircraft without needing to connect a USB cable.

Having a telemetry link isn't essential but it's definitely something I'd recommend, as it allows you to monitor your aircraft's position on a moving map, set waypoints and issue other commands for your aircraft as it's flying. It's also useful when you're not flying, at it saves the hassle of having to plug in a USB cable each time you need to check or change a parameter. This is all controlled via the ground station software.

Ground station software

The ground station software is what you use to control your drone from the ground, where the telemetry modules are used to send and receive commands between ground station and aircraft. Many autopilot systems include ground control software, which usually consists of some sort of map so that you can track the progress of your aircraft and set waypoints for it. Some ground station software is designed to run on your smartphone or tablet, which is great as it removes the need for you to carry a laptop around when you want to go flying. Often you'll find that the ground station software is what you'll also use to install or update your flight controller firmware, and to change various parameters of your drone's set-up.

Tracking antennae

In order to get more range out of your telemetry and video feed a high-gain antenna can be used. These focus the signal but are very direction-specific, so if they're not pointing directly at your aircraft the signal isn't very good. To aid with this some pilots use a tracking antenna, which is a stand consisting of two servos so that the antenna can rotate and pitch up or down to always be pointing at the aircraft. This system requires telemetry, so the position of the aircraft is sent back to the ground so that the tracking antenna knows what direction to point in.

BELOW **Planning an autonomous mapping mission. The drone is programmed to fly a grid pattern over the field and take photos at set distance intervals, to be used later to create an up-to-date aerial map. For the best results the grid pattern is angled so that the drone is flying into the wind one way, and with the wind in the other grid pattern direction.** *(Author)*

ABOVE A few different brands of R/C radio transmitters. *(Author)*

It should also be noted that throughout most of Europe it isn't legal to fly your aircraft beyond unaided visual line of sight, so the use of tracking antennae isn't that common, as you need to obtain special permission from the CAA (Civil Aviation Authority) to fly your aircraft beyond this distance. Standard off-the-shelf telemetry/video transmission equipment will usually have sufficient range out of the box for your needs, so under normal flying conditions you should never need to worry about losing the telemetry link.

Radio control

Although a drone can technically fly on its own, radio control (often abbreviated to R/C or RC) is a critical part of any UAV system as it provides a way to manually control your drone if required. A radio-control system consists of a transmitter (the device you hold in your hands) with sticks and buttons to control your aircraft. As you move the sticks or press buttons your R/C transmitter will send those commands to a receiver on board your aircraft via radio signals. The receiver will read the signals and convert them into outputs to your autopilot, which then performs the required action. In this section we'll look at some of the main features of your radio-control gear specific to flying drones.

R/C transmitter

A radio-control transmitter is the device that you usually hold in your hands when flying. Its two sticks are used for primary control of your aircraft by managing throttle, roll, pitch and yaw. There's also a variety of other buttons and switches that are used to control other functions, such as flaps, landing gear and autopilot-related controls such as changing flight modes and controlling the on-board camera.

Mode 1 or mode 2

R/C transmitters come in two possible modes, imaginatively called mode 1 and mode 2. These basically define the order of the primary control sticks. An easy way to visually tell the difference between them is to look for the throttle stick, which is easily identifiable as it isn't spring-loaded to return to the centre like the other sticks. If the throttle's on the left it's most likely

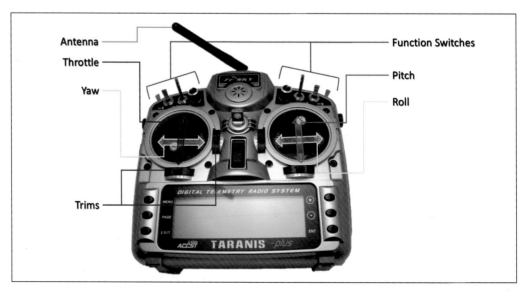

RIGHT An example of a typical mode 2 R/C transmitter. The two sticks are used for primary control of your aircraft, but there are several extra switches that can be used for special functions such as controlling an on-board camera, changing the flight mode on your autopilot and so on. *(Author)*

a mode 2 radio; and if it's on the right it's most likely a mode 1.

A mode 2 radio is the more popular setting. This has the throttle and yaw controlled by the left stick, with the up/down motion being throttle while yaw is controlled by the left/right motion. The right stick controls roll and pitch, where pitch is controlled via the up/down motion. Mode 1 radios, in which the throttle/yaw and roll/pitch stick positions are reversed, are more common in the Far East. However, choice of mode ultimately comes down to pilot preference and is adjustable with most radios, although this requires you to open up the unit and change some springs around. Consequently it's best to buy the mode you prefer from the outset.

Channels
When the word 'channels' is used in relation to radio-control transmitters it refers to the number of individual controls of which an R/C controller is capable, or outputs that its radio system has. Things such as throttle, roll, pitch and yaw are each one channel on the system. Typically when flying a drone you'll need to have at least five channels, so four can be used to control each axis of the drone (roll, pitch, yaw and throttle) while the final channel (usually a switch) is used to change the flight modes on your autopilot. However, it's more common to have more channels so that you can have added control for things such as tilting your camera up and down, taking a photo or raising and lowering the landing gear. You can therefore think of channels as representing each function that you can control on your R/C aircraft. With some fancier R/C systems you can actually control several things with a single channel, thereby enabling 'mixing', as discussed overleaf.

In most radio systems channel 1 corresponds to the roll stick on your R/C radio; channel 2 is for pitch; channel 3 is for throttle; and channel 4 is for yaw. But some R/C brands might be different, or you'll be able to set what each channel does in the radio settings. For more information on how to do this you'll need to check your R/C transmitter's manual.

Channel reversing
One of the most important things to check before flying is to make sure that each channel is set in the correct direction, so that if your throttle stick is down (*ie* set to minimum) your aircraft reads this signal as down too. I've seen too many aircraft crash because this step has been overlooked before its maiden flight. Depending on your radio control system or the autopilot that you're using it might read the R/C signals slightly differently, consequently all R/C transmitters have the ability to reverse a channel. This allows your aircraft to respond how you expect it to, so that when you move the roll stick all the way to the left your aircraft reads this correctly. This is particularly important when using servos on a fixed-wing aircraft, as depending on how they're mounted a maximum signal might cause it to move the control surface on the wing the opposite direction to what it should. Simply reversing that channel on your transmitter will solve the problem.

Model memory
Some more advanced radio transmitters have a built-in screen and menu functions that let you change the settings to alter how they work. As most pilots will use a single R/C transmitter for many different models, a useful function is model memory. This allows you to save a specific group of settings that you've configured for each vehicle, and then load it before flying the respective models. So you can use the same transmitter with a fixed-wing plane, multicopter and even a car as long as you've configured the various models correctly.

Dual rates and exponentials
Dual rates and exponentials adjust the outputs in relation to your stick inputs. To makes this easier to understand we'll consider only the roll stick (aileron control) on your R/C transmitter. With default settings we can assume the neutral position with the stick centred produces an output of 0. When we move it all the way to the left the output will be -100%, and all the way to the right will produce output of +100%. So the output response is linear in relation to your stick input.

Setting dual rates of 50% for low would mean that the outputs are effectively reduced by half, so if you move the stick all the way to the right the output would now be 50% less. Dual rates are particularly useful when you have a fast fixed-wing aircraft. As you fly faster the amount of movement you need on the control surfaces reduces, so switching to low dual rates makes the aircraft less sensitive and easier to control, as the control surfaces will only move to 50% of their normal range. The dual rate setting is controlled via a switch on your radio-control transmitter. However, with some of the advanced features of an autopilot dual rates aren't typically something that you'll use with your drone, but usually only in manual R/C aircraft.

Rather than a linear relationship between your stick inputs and output, it's possible to adjust this by means of the stick's exponential function. Exponentials have the advantage of still providing the full range of stick outputs while giving you more precise control around the neutral region; so if you move the stick all the way to the right you'll still get an output of +100%, but now this output is scaled so that when you move it, say, 25% to the right, the output will be scaled back to, say, 10%. This allows you to move the stick around the neutral position more, making it less sensitive, which is ideal for more acrobatic aircraft. I usually use around 30% exponential as a starting point and work from there. However, some miniature racing quadcopter pilots set the exponential to 100% to give the greatest possible control around the neutral regions in order to stop the quadcopter's flight from being too twitchy. But exponential setting is something that you'll need to play with until it feels right for you.

Mixing
Mixing allows you to control several outputs with a single channel, and is quite common with larger regular R/C models. A frequent use for mixing on aerobatic radio control aircraft is to mix the flap controls with roll control, so that when you move the roll stick to extreme positions the ailerons and flaps will move simultaneously on your aircraft, allowing it to roll faster than if you used the ailerons on their own. Another use on more basic radio-control aircraft might be so that, instead of needing to use a combination of aileron and rudder when you turn, by mixing the aileron and rudder controls you only have to worry about moving the aileron stick, and the radio system will apply movement to the rudder for you. However, mixing is almost never used on multicopters, as the flight controller takes care of this. Consequently I won't go into detail about it here.

If you want to learn more about mixing, the best place to look is in the instructions that come with your R/C gear. I should mention here that mixing often causes frustration amongst novice multicopter pilots. Logically it would seem to make sense to set your radio into helicopter mode for a multicopter, but this mode is for traditional helicopters, which make extensive use of mixing. Mixing on a multicopter ends up confusing the autopilot (since this does its own mixing), and results in very strange flight behaviour. So whenever flying any UAV make sure to set your radio to regular aircraft mode with no mixing enabled to start with, and then add any mixing you may require for special applications.

Failsafe
This is a crucial but often overlooked aspect of model aircraft flight that needs to be configured when setting up your R/C transmitter. A failsafe defines the behaviour of your aircraft when its receiver loses its link with your R/C transmitter. Although a rare occurrence it could happen if the transmitter runs out of battery power mid-flight, or you fly beyond your R/C's transmission range.

Depending on the type of aircraft you're flying you'll need to set the failsafe appropriately. In the case of manual R/C aircraft that are aerodynamically stable, and will naturally return to straight and level flight, you need to set up the failsafe to set all the controls to their neutral position, with a little throttle and slight yaw so that the aircraft will circle until you can re-establish the R/C link. However, in the case of flying a drone this will confuse the autopilot into thinking it still has a link with your transmitter. Most autopilots read the failsafe via the throttle channel, so in the event of losing the R/C signal you want to set the throttle channel to a very low value so that your autopilot can recognise this and then enter its own failsafe procedure, which could be to return home and land or to hover or circle its current position, depending on what you set on the autopilot.

Transmitter modules

Some R/C transmitters have the ability to change the actual transmitter module by simply plugging a new one in on the back. This is often beneficial as it allows you to upgrade or change to new systems when you require different functionality. Some modern equipment allows you to combine both telemetry information and radio control signals so that you can monitor various parameters aboard the aircraft. Other more serious first-person-view pilots prefer to use specialist long-range equipment to enable them to fly further away. However, depending on where you fly this could be unlawful, as you aren't supposed to fly your drone beyond visual line of sight.

R/C receiver

A radio-control receiver is a small module mounted aboard your aircraft that receives the signal commands from your R/C transmitter. The R/C receiver is connected directly to the input connectors on your autopilot, which also provides power to the receiver.

Channels

Just as with R/C transmitters, receivers have a specific number of channels associated with them where each channel represents an output or something that you can control. But this doesn't mean that you have to have exactly the same number of channels on your receiver and transmitter. If you're flying a very basic model aircraft that only has controls for roll, pitch, yaw and throttle you only need to use a four-channel receiver, which is cheaper and smaller than a 14-channel receiver.

Binding

Binding is the process of connecting your R/C receiver to a specific R/C transmitter, similar to how you pair a Bluetooth headset to your phone. The binding process will vary depending on the brand of R/C system you're using – some make use of a bind plug that you connect to your receiver when you first turn it on, so that it will connect to the nearest R/C transmitter; others have a bind button that you need to press to enter bind mode. Irrespective of bind method, the documentation of your R/C system will explain how this is done.

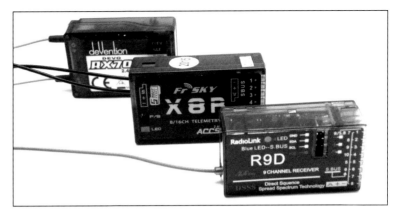

ABOVE Some examples of 2.4GHz radio receivers that can be used to receive the control signals from an R/C radio transmitter. *(Author)*

Antenna styles

One thing to consider when purchasing an R/C receiver is the type of antenna, as this gives an indication of the robustness and range capabilities of the connection. Most of the cheaper R/C systems make use of a single wire as their antenna. However, this often results in poor signal reception at longer ranges. An improvement on this would be to make use of two wires, known as a 'diversity antenna', which improves your ability to pick up signals because there are more antennae. Other receivers make use of plate/PCB antennae, which often have the best reception at greater ranges.

When you're mounting antennae on to your aircraft its important to mount them as far away from other electronics as possible. When using dual/diversity antennae it's recommended that they are spaced at 90° to one another to get

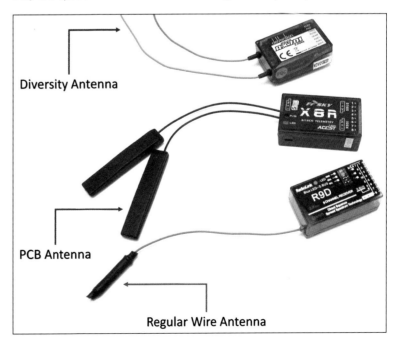

BELOW Examples of the three most common types of R/C receiver antennae. *(Author)*

the best possible signal range. Another thing to bear in mind on carbon fibre multicopters is that their carbon fibre plates block out most radio signals, so that for best reception it's best to mount the antennae on the side or bottom of the frame so that they have a clear line from your drone to your R/C transmitter.

Why 2.4GHz?

The 2.4GHz waveband is divided into specific frequencies spaced around it, known as channels, which shouldn't be confused with the channels of a radio transmitter or receiver. A waveband channel is a specific single frequency within the 2.4GHz band, such as 2.412GHz or 2.422GHz.

Today you'll find that just about all R/C systems run on the 2.4GHz frequency band, as these have improved reliability, lower power consumption and increased robustness compared to the older 35MHz systems. In the old days you needed to be very careful when flying your aircraft in the vicinity of other R/C pilots as it was easy to accidentally use the same frequency/channel, which would interfere with each other's signals and usually ended in a crash. 2.4GHz systems don't have the same interference issues, as each receiver is paired or bound with a transmitter, allowing many pilots to fly in the same area without worrying about interference from other radios, thanks to spread spectrum and frequency-hopping technology.

Frequency hopping, simply put, means that your radio gear constantly hops from channel to channel to find the one with the least traffic; your radio gear will typically do this a thousand times a second. Spread spectrum, equally simply put, means that the information is spread in a very specific sequence of bursts within a given channel, so that the receiver knows exactly what to look for. This makes 2.4GHz radio systems very resistant to interference. Nevertheless, there are always other sources of potential interference, particularly as you fly further away and the signal gets weaker. Therefore care should always be taken to mount the receiver on your aircraft as far as possible from other transmission sources (such as your video transmitter) and electronic devices (such as motors).

PWM and PPM

PWM and PPM refer to methods by which the R/C signal is encoded. Without going into any unnecessary details, it's important to understand the difference between them.

PWM, which stands for pulse-width modulation, is the most common method used to manage items such as electronic speed controllers (which control the motors) and servos. Most R/C accessories, such as ESCs and servos, use this method of communication. PWM requires a single wire per channel, so if you're using an eight-channel receiver you'll need to use eight connectors. This can sometimes make your drone look a bit messy and can get confusing when you're connecting everything, as you need to connect each channel to the correct input port on your autopilot.

PPM, or pulse-position modulation, can be thought of as lining up several PWM signals one after another. The most noticeable difference of PPM is that a single wire can carry several channels' information, so you can connect an eight-channel receiver to your autopilot with just a single connector, which makes it very easy when it comes to connecting things.

Neither PWM nor PPM is better than the other, although many drone builders prefer PPM as it tends to facilitate a neater and easier connection between your receiver and autopilot (only one wire required). On the other hand PWM has some advantages of its own, as you

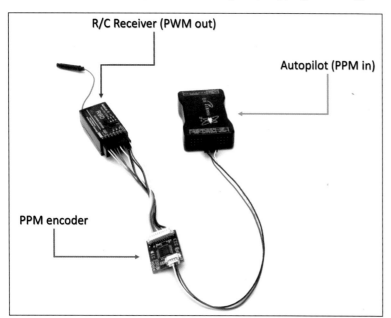

BELOW A PPM encoder will convert the signals (one wire per channel) to PPM signals (a single wire for all channels), so that the autopilot only requires one input. Some R/C receivers have the ability to output PPM directly so you don't need to use a PPM encoder.
(Author)

can bypass the autopilot directly with some channels if you need to control specialist gear or functions on your drone.

There are also other brand-specific digital proprietary protocols available, such as SBUS, or DSM, which are similar to PPM in the sense that you only need one cable to connect with your receiver.

PPM encoder

Some autopilot systems have only a single PPM input, which restricts their compatibility as you can only use receivers that output PPM; but it's still possible to use a regular PWM receiver with a PPM-only autopilot by using a PPM encoder. A PPM encoder will combine all PWM connections and output them into a single PPM output that you can plug into your autopilot.

Motors

The primary types of engine used for radio-control aircraft today are brushless electric motors, as these provide a substantial amount of power given their low weight. Simply put, a brushless motor consists of a stator (non-moving part) and a rotor (the rotating part). The stator consists of coils arranged in a radial pattern, with copper wire wound around each coil, to form a bunch of electromagnets. The electromagnets/coils are wound and connected in a very specific way depending on the desired characteristics of the motor.

The rotor consists of magnets arranged around the inside of the outer motor casing (also known as the bell). In order to cause the motor to rotate, power is applied to specific sets of coils at very precise intervals, which is why you need to use an electronic motor controller to govern its speed. This also explains why brushless motors have three wires to connect them, because there are three sets of connected electromagnets in a typical brushless motor.

In terms of knowing what motor you should choose for a given drone there's no single answer, as it depends on many factors such as prop size, required lift and whether or not you want acrobatic capability or would prefer a longer flight time. More details on choosing the correct motors are discussed in the 'Choosing

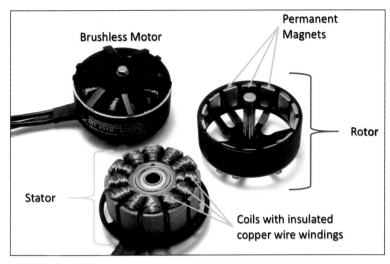

components for your drone' section, which covers some of the basic technical details of drones.

Outrunner and inrunner motors

There are two main categories of brushless motors. The outline description provided above is actually for an outrunner motor, which is the type used almost exclusively when it comes to either electric multirotor or fixed-wing UAVs. Outrunner motors have their rotating part on the outside, hence their name. By contrast, an inrunner has its rotating part on the inside of the motor while the outer shell remains stationary. With inrunner motors the permanent magnets are attached to the shaft and the electromagnets are mounted on the inside of the outer shell – the exact opposite of an outrunner.

If outrunner motors are used almost exclusively for drones you might wonder why you need to know about inrunners. The quick answer is that inrunner motors are more commonly used with radio-control cars. In order to understand the reason for this we'll need to go back to some senior school physics and recall the principle of torque.

The fact that the heavy magnets on the rotor of an inrunner motor are very close to the centre means that they don't produce much torque. This allows inrunner motors to spin much faster, making them suitable for a few specialist aircraft applications such as electric ducted fan engines, which have very short propellers that need to spin very fast to produce the required thrust. It's this high rpm capability

ABOVE Top left: a typical multirotor (pancake-style) outrunner motor. Right: the rotor has been removed from the brushless motor and turned upside down, showing the permanent magnets arranged around its inside. Bottom left: the stator of the motor has coils arranged in a radial pattern; electrical current passes through these at specific intervals to cause the rotor to spin. *(Author)*

RIGHT A typical motor naming convention: the first four numbers are an indication of motor size, the numbers followed by KV show how fast the motor will spin, and the final numbers provide details of the stator and rotor configurations.

2212Q 850KV 12N14P

- 22 / 12 — Motor or Rotor Diameter / Motor or Rotor Height
- 850KV — Revolutions per Volt
- 12N — Number of Electromagnets in Stator
- 14P — Number of Permanent Magnets in Rotor

that makes inrunner motors a popular choice on radio-control cars. With outrunner motors, because the moving parts are located further apart they're able to produce more torque, which allows you to use much larger propellers on them, making them ideal for use with multicopters and planes.

Typically you'll find that motors able to drive very big propellers have quite a large diameter so that they can produce enough torque. So a motor designed to drive a 10in propeller will usually have a smaller diameter compared to a motor designed to drive a 15in propeller.

Motor size

Typically when you buy brushless motors the model numbers are defined by a series of four digits sometimes followed by another two. Most of the time these numbers aren't just arbitrary values, but rather define certain characteristics of the motor. Unfortunately there's no universal standard in motor-naming conventions, but in general the first four numbers are an indication of the motor's physical size. The first two indicate either the entire motor diameter or the stator diameter. The second two define either the entire motor height or the stator height, depending on the manufacturer.

Motor KV

The other important factor in motor numbers is the KV rating, which represents the rpm of the motor per volt – so an 850KV motor will rotate with an rpm of 850 if it has 1V of energy applied to it. A KV mustn't be confused with a kV (kilovolt), which isn't the same thing.

As you'd imagine, lower KV motors spin slower (usually have more torque), and higher KV motors spin faster. A low KV motor is created by using windings of thinner wire around each electromagnet; the motor therefore carries more volts for fewer amps, which means it spins slower but produces more power/torque. A high KV motor is created by using thicker wire with fewer windings around each electromagnet, meaning it carries more amps at lower voltages. This allows the motor to spin faster but at a lower torque, making high KV motors better suited to smaller props.

Motor configuration

The configuration number tells you how many electromagnets there are on the stator, and the number of permanent magnets on the rotor. The number before the letter N shows the number of electromagnets in the stator, while the number before the P shows how many permanent magnets there are in the rotor. Most outrunner brushless motors follow the 12N14P configuration. There are some specialist low-KV multirotor motors that have a higher number of electromagnets and permanent magnets, which allow the motor to create more torque more efficiently. However, due to the extra magnets these are more expensive.

BELOW Brushless motors come in a range of sizes based on application. The bigger the motor, the more powerful it is; the larger its diameter, the slower it will usually spin. *(Author)*

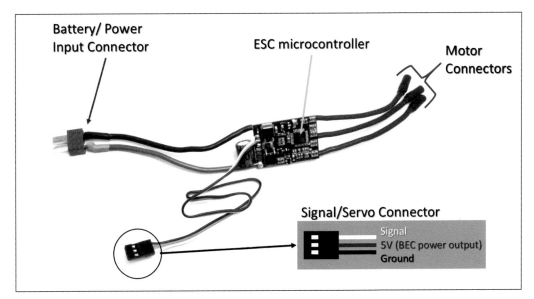

LEFT An ESC with its covering removed, showing the internal electronic components. An ESC consists of three motor connectors, a power input connector that provides power to the ESC, and a servo connector that reads the signal from your autopilot. If the ESC has a built-in BEC then this servo wire will also include a 5V power output that can be used to power other electronics, such as your autopilot. *(Author)*

Electronic speed controllers

ESCs, or electronic speed controllers, are devices used to control brushless motors. A brushless ESC sends power to the motor at specific timings to cause it to rotate at a specific speed that you choose. As explained in the 'Motors' section a motor has three wires sticking out, where each of those wires is connected to a specific number of stators located around the motor. Each of these is called a phase. The ESC sends signals to a specific section (phase) of the motor to activate the electromagnets at very specific timings, to cause it to spin. This is all controlled by a little computer (called a microcontroller) that's located inside each ESC.

ESC size

When we talk about ESC size we're not usually referring to its physical size, but rather the size of its output in amperes. The size of an ESC will define the size of motor that it can drive, with 10–12A ESCs being common for mini quadcopters, 20–30A being common on quadcopters that carry GoPro-size cameras, and so on. When choosing an ESC it's important to know what motor and propeller you're using with it, so that you can match it all up to ensure that your ESC can provide sufficient power, otherwise you could end up burning it. Full details on how to go about choosing motors, ESCs and propellers is discussed later, in the 'Choosing components for your drone' section.

BEC (battery eliminating circuit)

Some ESCs have what's known as a BEC, or battery eliminating circuit. All that this means is that ESCs with a BEC are able to output constant voltage that you can use to power your R/C gear, such as receiver, servos and flight controller. A BEC is often not required with most autopilot-based drones, as the autopilot itself has a power module that serves the same function. But when you're powering a large number of servos or other specialist gear that draws a lot of power it's important to check the specifications of the BEC, as this will often have a maximum amount of current draw. If your

BELOW A variety of ESCs of various ratings, from a small 10A ESC used on mini quadcopters (front) all the way to a large 60A ESC used to drive very large motors. *(Author)*

equipment uses more current than the BEC can provide you're likely to crash your drone. Using too many servos that are activated at the same time would draw the most current. However, the majority of BECs are able to handle at least four, so for most drones you don't need to worry about this. Just check the specifications of your servos to see the current draw and make sure when you add them all up that it's not more than what's stated on your BEC current output.

Most of the cheaper ESCs use a linear BEC that steps down the battery voltage to 5V by dissipating the excess energy as heat. To avoid this, some use SBECs (switching BECs) that don't have the same heat dissipation issue. However, because of the switching nature of the ESC it can generate extra interference, which can be picked up by other R/C gear connected to it, such as the autopilot or R/C receiver. For this reason when using SBEC ESCs it's best to remove the red wire from the servo connector as discussed later in this section.

To avoid any possible interference conflicts some ESCs are actually isolated from the main power circuit by means of an optical isolator, to ensure noise from the motor circuit won't carry through to your other equipment. Often referred to as OPTO ESCs, these don't have a BEC built into them and don't output any power to the autopilot or R/C gear. They're particularly common on multicopters because they have more than one ESC connected to the flight controller – some ESCs with BEC built in can interfere with each other, as there's more than one ESC outputting power to the flight controller. If you do only have ESCs with BEC then a common practice is to disconnect all the power supply cables from them except one, so that only one ESC is providing power to the flight controller or R/C gear.

Removing the power cable from an ESC with BEC (multirotor)

Some multirotor flight controllers and ESCs don't require you to remove the 5V power cable from your ESC, as having multiple ESCs feeding power to the flight controller won't cause any interference issues. However, it's common best practice to do this anyway, to avoid any possible conflicts, as removing the connector is very quick and easy. Once that's done you need to insulate the end of the red wire so it doesn't come into contact with anything. This can be done with some heat shrink or electrical tape. Some people simply cut the red wire, but I don't recommend doing that in case you need to use the ESC on another project in the future. The method outlined here means that if you need to use the 5V power line you can simply clip it back into the ESC servo connector.

RIGHT An example of four quadcopter ESCs connected to a flight controller. The power cable has been removed from all but one of them. This is to avoid any chance of interference. One ESC is still needed to provide power for the flight controller. *(Author)*

1 Use a pair of tweezers or something else sharp to lift the plastic clamp...

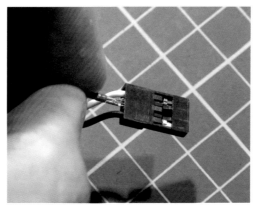

2 ...then slide the red power cable out of the servo connector.

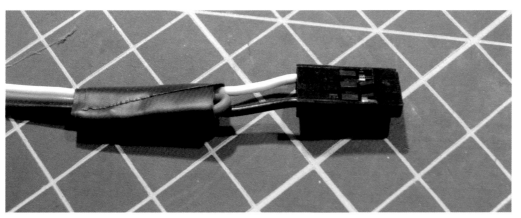

3 To avoid any chance of the wire causing a short circuit or touching any other part of the electronics, simply insulate it with some electrical tape. *(All photos Author)*

Multirotor ESC firmware

Fixed-wing drones are fairly traditional in terms of motor control, as electric R/C aircraft have been around for many years, so using any sort of brushless ESC brand will work just fine. However, when it comes to multirotor drones these have additional requirements because their ESCs need to change speed very quickly, as varying the speed of the motors is how you control a multicopter in the air. Consequently custom firmware has been written specifically for multirotor aircraft applications. SimonK (written by Simon Kirby) and BLheli (written by Steffen Skaug) are the two main types of multicopter firmware. Both started development around the same time, but the SimonK firmware was written for ESCs using Atmel microcontrollers, while BLheli was written for Sllabs microcontrollers. You might think of Atmel and Sllabs microcontrollers as being like Windows and Mac, where some Windows software isn't compatible with Mac etc, but recently both firmware versions have been made to work on all the main ESC microcontrollers.

The great thing about these multirotor firmwares is that the authors have shared and published the source code, so anyone can use and modify them should they want to. While there are many discussions on the Internet about which firmware is better, at the end of the day they're both, in their latest versions, designed to do the same thing, and both work well on multicopters.

In the past the SimonK firmware had some issues with low KV motors, but in recent versions this has been fixed. The BLheli firmware has the ability to configure settings via PC software connected to USB devices, which does have some advantages over using the

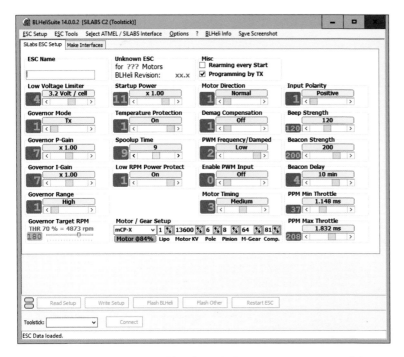

ABOVE Screenshot of the BLheli PC software that you can use to configure your ESC settings.
(Author)

traditional musical tone set-up process. (More on ESC programming later in this section.)

The main difference between multicopter ESCs and regular ESCs is that with multirotors the update commands are sent much faster to the motor, giving your quadcopter much more precise and accurate control. With traditional ESCs the inputs are sometimes smoothed out so that your motor will speed up smoothly, which is useful as it prevents damage to gears in some helicopters or older-style fixed-wing aircraft. It's still possible to use regular ESCs with your multirotor, its just that it might be slightly sluggish.

Another distinction is that by default multirotor ESCs don't have any low voltage cut-off. What this means is that when your flight battery voltage gets low as you're flying, traditional ESCs have a safety feature that'll turn off the motors but continue to provide voltage via the BEC so that you can still control the aircraft's servos. This is great for fixed-wing aircraft so that you can glide your plane down to safety if the battery runs flat. However, in the case of multirotors you don't exactly want your motors cutting out all of a sudden! So for many multirotor ESCs this feature is disabled so that you can still bring your drone down to the ground safely. Most flight controllers are able to monitor battery voltage to warn you when it's dropping so that you can land before it's too low, as discharging your battery below the suggested voltage can damage it.

Programming

All ESCs have a range of parameters that you can change according to aircraft type. The most common (and slowest) method is to use the musical-tone programming menu, which uses audio tones to change various parameters. A slightly easier method is to use an ESC programming card that lets you set up all the settings you want and then load them on to your ESC automatically in one go – especially useful for multirotors when you have more than one ESC to program. The last method is to connect your ESC to your computer and program it via software, known as USB programming.

A safety warning: *always* make sure you remove all the propellers from your aircraft before doing any sort of ESC programming or calibration, as you don't want the propellers spinning up accidentally and causing injury to you or damage to nearby equipment.

Musical-tone programming

To enter the musical-tone programming menu you need to send a high signal command (usually by moving your throttle stick to maximum) before you plug your battery into the ESC. When your ESC enters programming mode it will play a series of beeps that will change depending on what menu section you're in. The actual tones are usually specific to the manufacturer of the ESC, but a table is always provided in the manual to show the various beep sequences and what they mean.

Programming card

A programming card is a device on which you can set up all the parameters visually on an electronic card that you then plug into your ESC to set them all in one go, making this a much faster way to program. Programming cards often have LEDs that will light up depending on what setting you've selected. Once you've gone through all the settings you just need to hit the 'OK' button to save them on to the ESC.

USB connection

This method involves connecting your ESC to your computer via a USB cable, to change the settings via computer software. This is my preferred method as it allows you to set up everything exactly how you want it, and ESCs that

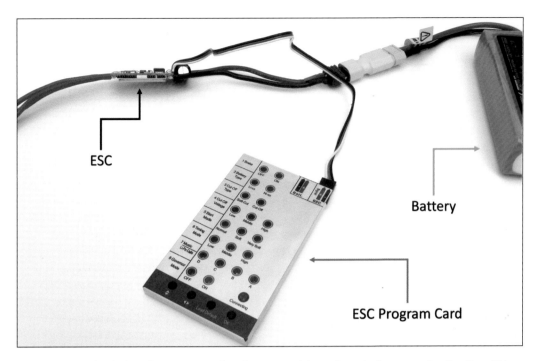

LEFT Using an ESC program card can make programming an ESC much faster if it can't connect to your PC via USB. *(Author)*

support this method often allow you to update the firmware as new versions are released. If using this on Windows you usually need to install specific software drivers in order to get the ESC to talk to your PC, but details on how to do this are usually included with the manual.

Common ESC settings

In this section I'll go over some of the main ESC settings and explain what each does. However, since most manuals provide some details, rather than repeating what's in them here's a brief definition of what each setting is used for.

Voltage cut-off

As discussed above, when the battery voltage gets low, some ESCs will stop sending power to the motor in order to preserve power for your other electronics, such as the R/C receiver, flight controller and servos. This is ideal for fixed-wing aircraft, but not for multirotor drones. For fixed-wing aircraft, set this to hard cut-off so that the motor will stop immediately when the battery voltage gets too low. For multirotors the best option is to set it to soft or super soft, which will gradually reduce the power output to the motors so that your aircraft will descend gently to the ground.

Braking

When no throttle command is sent to the ESC it will force the motor to stop spinning by applying a force in the opposite direction. This is sometimes great for fixed-wing planes with foldable propellers, as it means they can fold away for more efficient gliding. By default this is often disabled on the ESC. For multirotors it's best to keep it off unless the given ESC/flight controller supports active braking, which can provide increased stability to your quadcopter.

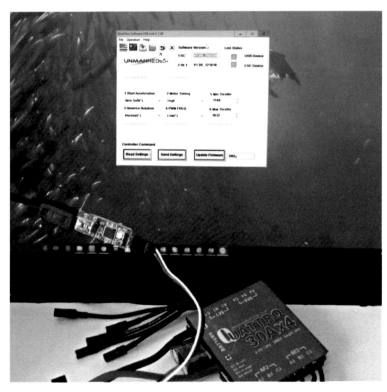

BELOW Programming a 4-in-1 ESC via a USB programming tool allows you to use software to select all the settings. *(Author)*

Timing

This has to do with the timing at which the ESC sends power to each of the three phases of the motor to cause it to spin. When the timing isn't set correctly you might notice the motor will oscillate as it tries to spin first one way and then the other due to poor timing. This can also happen when you change the throttle quickly, when the motor might behave strangely if the timing is set incorrectly. This isn't something you often need to change unless you're using certain low KV motors (less than 500KV) or high pole count motors as used on some multirotors, which typically need higher ESC timing values to spin properly. If you do change the timing values you should properly test that the motor doesn't act strangely at any particular throttle value or when changing throttle value from max to min very quickly.

Switching frequency

This is typically set by default to be either 8kHz or 12kHz. Increasing this setting can result in the motor running more efficiently but will cause your ESC to generate more heat, as the current is switching much faster inside it. However, as a rule of thumb you can work out the best approximate switching frequency by the following equation, based on the number of poles in your motor and its KV rating:

$$\text{Switching frequency} = \frac{\left\{\frac{\text{number of poles in motor}}{\text{motor KV} \times \text{battery voltage}}\right\}}{20}$$

Again, if you're increasing switching frequency it's a good idea to run your motors with propellers on the ground for about five minutes before you go out to fly, just to check that the ESCs or motors aren't getting too hot.

Start-up acceleration

The start-up acceleration, as you might have guessed by its name, has to do with how quickly the motor will spin to the throttle you've set. For multirotors you'll want this to be set to 'hard' – the highest setting – to make sure the motor responds as fast as possible. For some fixed-wing aircraft with large motors you might want to set some sort of acceleration delay to cause the motor to change its speed more smoothly, as sometimes the initial torque from the motor changing speed quickly can cause your aircraft to roll. Also, this setting is useful if you're using any gears between your motor and propeller, as the softer acceleration will increase the life of the gears and reduce the risk of the gear teeth breaking.

Calibration

An important step during set-up is to calibrate your ESC with your autopilot. The process of calibration tells your ESC the maximum range of input signal values and how they should correspond to the throttle. This process is explained in the diagram below left. Here you'll see a bar with two braces; the top brace represents the ESC throttle range and the bottom brace represents the input signal from your R/C transmitter or autopilot. By default an ESC arrives from the factory with some arbitrary throttle range, but because each autopilot or R/C transmitter has a slight variation of output signal values the two might not match up, as you can see in the illustration of an uncalibrated ESC. When your ESC is uncalibrated you'll notice that when your throttle input is at maximum (about 1,900) your ESC won't detect this as maximum, as in the illustration the maximum value for the ESC is at 2,000.

The process of calibration simply moves the ESC throttle range to exactly match the input, so that when your input throttle is at maximum so is the ESC throttle output to the motor. As you can see in the calibrated diagram, the two ranges match up perfectly.

BELOW Top: illustration of an uncalibrated ESC where you can see that the throttle range of the ESC doesn't match the throttle range of the TX/autopilot. Bottom: the calibration process makes sure that the throttle ranges are matched up to be the same for both the input (TX/autopilot) and the output to the motors. *(Author)*

How to calibrate your ESC

The calibration process is very easy and straightforward. The most basic, but most time-consuming method is to do a manual calibration one ESC at a time. The preferred and faster method, particularly for multirotors, is to calibrate all your ESCs at once via your autopilot software or an ESC hub. Although the process might alter slightly between different autopilots and ESCs the basic process is outlined as follows. For full details see the documentation provided with your specific ESC and autopilot.

Remember, you should *always* make sure that you remove all the propellers from your aircraft before doing an ESC calibration, as you don't want the propellers spinning up accidentally and causing injury to you or damage to nearby equipment.

The first step is to turn on your radio transmitter and move the throttle stick to the maximum position, then turn on or plug your battery into your ESC. The ESC will start playing the start-up musical tone followed by another beep (or two beeps depending on the ESC). At this point your ESC will record the maximum throttle value from your transmitter. After the two beeps move the throttle stick back down to the minimum value. Your ESC will then record the minimum throttle value and indicate it's saved it by a single long beep. Now you can test it's all working as it should by moving the throttle stick slightly up, when your motor should start spinning.

An important check with multicopters is to ensure that all the ESCs have the same calibration settings. If correctly calibrated, all motors will start to spin at the same time when on level ground. If they don't then there's probably an issue with the ESC calibration and you should perform the calibration again.

Reversing the direction of a motor

Depending on your drone, it might be necessary to reverse a motor's spin direction. This is particularly important on multirotors, as the spin direction of the motors controls the drone, as defined by your autopilot documentation. When connecting an ESC to a motor, depending on the order of the wiring your motor will spin in a particular direction. If the motor spins in the wrong direction simply swap the two outer wires to reverse it. Also, some ESCs have the ability to reverse motor direction via their settings, which is very convenient – particularly if the motor/ESC wires are hard to access.

With brushless motors there's no particular order in which you need to connect the motors and ESC together – you can't damage anything if you plug the ESC/motor wires in the wrong way

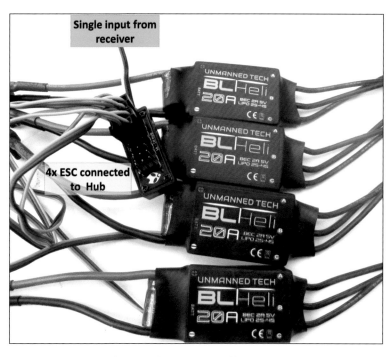

ABOVE Calibrating your ESCs with an ESC hub makes the calibration much faster, as you only need to do it once and the calibration is applied to all ESCs that are connected. *(Author)*

BELOW The top ESC/motor has the two outer wires crossed over, which will cause the motor to spin anticlockwise. The bottom motor doesn't cross any ESC wires so will spin clockwise. *(Author)*

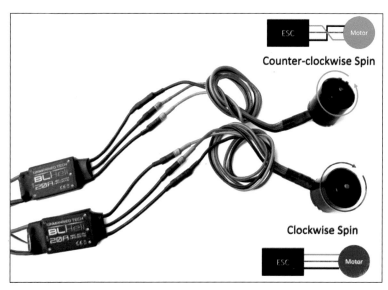

round. The order in which they're connected will define the motor spin direction. So if your motor/ESC doesn't have colour-coded wires and your motor is spinning in the wrong direction, just swap the two outer wires to reverse the spin.

Propellers

Propellers create the thrust to drive your aircraft, enabling it to fly. When it comes to choosing the propellers for a drone their size is largely dependent on what size motor and frame you're using. As a rule of thumb, it's usually best to use the largest possible propeller you can on your aircraft to maximise efficiency. If your propeller is bigger it can produce more lift at a given rotational speed; a smaller propeller spinning at the same speed will produce less thrust. This means that you need to spin larger propellers more slowly. When your propellers are spinning slower they're more efficient, because the air has more time to stabilise as the propeller spins around, making the air hitting its leading edge less turbulent.

Materials

The material from which a propeller is made is an important factor to consider, as it dictates its flexibility, cost and strength.

Plastic

Plastic is the most common material used due to its low cost. A range of plastic composite types is used in propellers, from nylon-based props – which are the most flexible – all the way to ABS (acrylonitrile-butadiene-styrene), which is a bit more rigid. Having flexibility in your props isn't always a bad thing as it makes them harder to break, since they're less brittle. Also, this flexibility can to some extent absorb some of the aircraft's motion, making it seem more stable and less jittery. The main drawback of plastic propellers is that they aren't as strong as other materials, making them unsuitable for heavy-lift platforms. Their flexibility can also cause props to twist and distort while spinning, particularly under load, making them less efficient.

Reinforced plastic

Reinforced plastic props are a good compromise between plastic and carbon fibre, making them quite strong but still quite cheap. These reinforced propellers are often manufactured using a plastic and carbon composite that makes for a more rigid propeller when compared to plastic. This extra rigidity means that they keep their shape better than pure plastic propellers and are a bit more efficient.

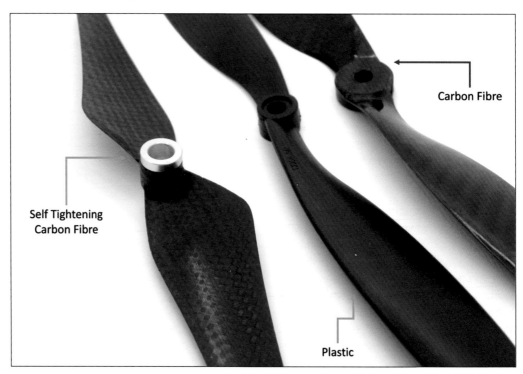

RIGHT The two main materials used on propellers are plastic and carbon fibre.
(Author)

Carbon fibre

These propellers are the most rigid, as carbon fibre is both very strong and lightweight. Due to the higher cost of carbon fibre props, however, they're less suitable for starter pilots. Nevertheless, their rigidity makes them the most efficient aerodynamically as they keep their shape better than other props. Furthermore due to their rigidity these propellers are better suited to heavier drones that generate more thrust. It's important to also consider the safety aspects when using carbon fibre: because carbon fibre is much more rigid than plastic, if you get your fingers caught between the blades it can cause much more damage, so I'd only recommend using these propellers when you're a more experienced pilot.

Wood

Wooden propellers aren't very common these days due to their expense. As they're made from wood they're also much heavier when compared to plastic or carbon fibre. This added weight means that they have more momentum as they spin, making wooden props unsuitable for acrobatic-style multirotors. Such propellers are sometimes used on heavy-lift filming drones.

Size (diameter and pitch)

When it comes to buying a propeller you'll find that they're most often sized according to their diameter and pitch, represented by four numbers, usually in inches. The first two indicate the diameter (distance between each tip of the blade) and the second two represent the pitch, which is the measure of how much air the propeller will push or how far it'll travel forwards in a complete rotation. The pitch gives an indication of how steep an angle the propeller blade has, but it isn't its actual angle.

Another aspect is the number of blades the propeller has. This is most commonly two, but sometimes you'll see three-bladed propellers. In terms of efficiency two-bladed props are better; however, on certain aircraft you're limited by the maximum propeller diameter, so having three blades instead of two in the same diameter means you can produce more thrust at a given rotational speed. Three-bladed propellers are more common on some fixed-wing aircraft and mini quadcopters due to the practical limits on maximum propeller diameter.

Prop Diameter and Pitch Marker

ABOVE A selection of plastic propellers in various colours. Most plastic propellers include an indentation that specifies the size and pitch, usually in inches. *(Author)*

In terms of choosing the correct propeller for a motor, this information is included within the motor's specifications, as each motor is designed to be used with a certain range of propeller sizes. Using a propeller that's too large can strain your motor, causing it to get very hot during use, which can damage the magnets and bearings within it. Consequently to get the longest life out of a motor you should only use propellers that fall within its recommended range.

Foldable propellers

As their name implies, foldable propellers have the ability to fold back during flight or for storage. This is particularly common on fixed-wing glider aircraft that have a motor – once you've flown to a high enough altitude you can turn off the motor and the propeller will fold backwards due to wind resistance and thus become more streamlined, which aids the aircraft's gliding efficiency. As the motor spins up again the centrifugal force will cause the propeller to unfold.

On a multirotor, rather than folding backwards the propellers fold sideways to make it easier to store and transport. Again, as the motors spin up the centrifugal forces cause the propeller to unfold in order to generate lift. The only main disadvantage of using foldable propellers on a multirotor is that they tend to be more expensive and are heavier than fixed propellers.

RIGHT Most shaft-mounted propellers include a selection of spacers that ensure a snug fit between motor and propeller. *(Author)*

Fastening your propeller

When it comes to mounting propellers to motors you again have several options, though the brand or style of motor you use will often define what propeller fastening methods you can use. When you purchase a propeller you'll often receive a selection of plastic rings of various sizes that you use to make sure the propeller fits snugly over the shaft. This ensures that it's centred perfectly to avoid vibrations and balance issues. Examples of such rings are shown here.

You have to select the right ring to fit snugly on to your propeller adapter and motor shaft.

Collet

The collet is a popular propeller mounting method for fixed-wing aircraft, particularly with pusher motors. The propeller is attached to the collet, which slides over the motor shaft and is secured via a grub screw. The great feature of a collet is that during a crash it can slide easily, causing the propeller to fall off the shaft. This helps prevent damage to the propeller and motor. However, because of this it isn't great for using on an acrobatic multirotor, since sudden changes in motor velocity could cause it to come loose over time, resulting in the propeller flying off mid-flight. That's not to say you can't use this method on a quadcopter, it's just not recommended.

Direct mount

Direct mount propellers are mainly used on larger multirotor platforms, particularly ones that use carbon fibre props. Having

RIGHT A propeller mounted with a prop collet. *(Author)*

RIGHT A propeller mounted directly to the motor (left) and a propeller mounted using a prop adapter (right). *(Author)*

your propeller mounted directly on top of the motor means that the motor's centre of gravity is lower and that there's a more secure connection between it and the propeller, which reduces vibrations. However, the disadvantage of direct mounting is that the propellers are more likely to break in a crash, as the propeller can't slip around the shaft as with other mounting methods.

Prop adapter

This is probably the most common mounting method currently used. The motor has a threaded shaft that the propeller slides on to, where it's secured by a large nut. When it comes to multirotor motors you can also choose the direction of the thread to avoid the nuts coming loose during flight.

Self-tightening

This method is similar to a prop adapter, but the bolt is built directly into the propeller. On the motor there's simply a threaded shaft on to which the propeller screws. Due to Newton's second law, the thread of the shaft is opposite to the rotation of the motor, so that when the motor spins the propeller automatically tightens on to it. This method is seen on many ready-to-fly quadcopter models, and it's very effective at preventing propellers coming off during flight because you forgot to secure them properly! Another useful attribute is that when you're done flying it's very easy to take the propellers off without any tools to avoid breaking or damaging them in transit.

Balancing your propeller

This is a particularly important step if you're using your drone for photography purposes, as unbalanced propellers are the main cause of the unwanted vibrations that cause photos to turn out blurry. Vibrations may also result in performance issues with the sensors in your autopilot.

The process of balancing, as the name suggests, minimises vibration by ensuring that both sides of the propeller are exactly the same weight, so that the centre of gravity is in the centre of the props and motors. A propeller balancer is essentially a shaft, on which the propeller is mounted. It has very little rotation resistance (it's usually suspended between magnets), so the propeller is free to rotate even if it's slightly unbalanced. You can think of the propeller balancer as being like a see-saw: if one side of the propeller is heavier, it'll drop lower than the other side; when both sides have equal weight the propeller is perfectly balanced and will stay level. There are two steps involved when it comes to balancing your gimbal. The first is to balance the blades, and the second – often forgotten – is to balance the central hub of the propeller. Details on how to balance a propeller are given below.

Balancing your propeller blades

The first and most important step is to make sure both sides of the propeller blades are balanced.

LEFT Self-tightening propellers are easy to remove without any tools. *(Author)*

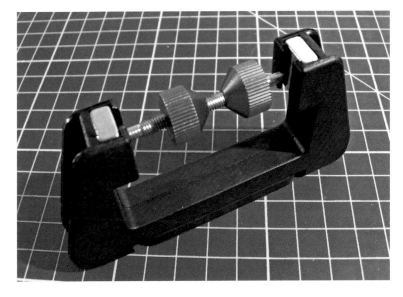

BELOW Example of a typical prop balancer that has a shaft suspended between two magnets, which allow for an ultra-low resistance contact that's free to rotate to assist the propeller-balancing process. *(Author)*

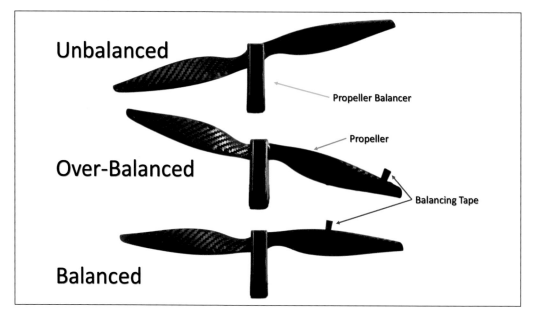

RIGHT Top: an unbalanced propeller – the left side is slightly heavier, and consequently falls downwards. Middle: a piece of tape has been added to the lighter side to weigh it down, which has instead caused the right side to become heavier. Bottom: the tape has been moved closer to the centre of the propeller to reduce its moment of inertia; now the propeller stays level, which means the blades are balanced. *(Author)*

1 With the propeller mounted on to the balance unit, the heavier side will droop downwards as shown. Now that we know which side is the heavier we need to add some counterweight to the other, to balance it out. In this example I'll use some tape, as it's the fastest and easiest method for balancing props. However, there are other methods, which we'll discuss later.

2 With a small piece of tape cut, stick it on to the outer edge of the lighter blade and move the propeller to the level position again.

3 If the blade then droops on the side to which the tape is attached, this means that the tape is too heavy. However, instead of cutting it smaller simply move it closer to the centre until the propeller stays level.

4 If the propeller droops on the opposite side you need to substitute a larger piece of tape or add an extra piece, to increase the weight of the lighter side. Once the propeller stays level you can secure the tape properly by firmly pressing it down.

I prefer to use tape as it's a quick and effective method of balancing propellers, but other methods include sanding down the heavier side to balance the weight (the best method for wooden propellers) and adding superglue to the edge of the lighter blade, though this can get messy very quickly!

RIGHT Some black electrical tape is folded over the leading edge of the propeller to balance the weight of the two sides. *(Author)*

Balancing your propeller hub

This step is often overlooked as it's not always necessary, since vibrations caused by an unbalanced hub are less noticeable than those of unbalanced blades. But this step is still useful to know about if your aircraft is having vibration issues.

1 Balancing the hub is a trickier process than the blades and requires some patience. First place the propeller on the balance at a slight angle, say 45°; if when you let it go it moves slightly to one side, flip the entire propeller around on the balance and try again.

2 After flipping the propeller and testing at various angles you should be able to feel if one side of the hub is heavier than the other.

3 Either add some tape or sand away some material to get both sides balanced. Once the propeller is perfectly balanced it should remain at whatever angle you place it.

Batteries

Batteries are what you use to power everything on your UAV, including motors, autopilot and other gear. The most common battery type used in today's drones is lithium-polymer (LiPo), which provides the best energy density compared to other batteries such as nickel cadmium (NiCad) or nickel-metal hydride (NiMH), which were previously common in the R/C hobby. LiPo batteries' high energy density means that for their weight, LiPo batteries store the greatest amount of electrical power. You'll also find lithium-based batteries in things such as your mobile phone, while laptops are commonly powered by lithium-ion batteries, which are similar to lithium-polymer but have a lower energy density, though they're slightly more rugged. However, lithium-polymer batteries do require special care to get the most life out of them (see below), and need to be stored safely. When it comes to selecting a battery, understanding these parameters will help.

For more details on how to choose a battery please refer to the 'Choosing components for your drone' section.

All batteries include two connectors, the main one with the thicker wires being the main power connector from which all the battery's power flows. The second smaller, multi-wire cable is called the balance lead. LiPo batteries consist of one or more cells, and each wire on the balance lead is connected directly to each cell in the battery; it's used for charging purposes and to ensure that each cell is at the correct charge level.

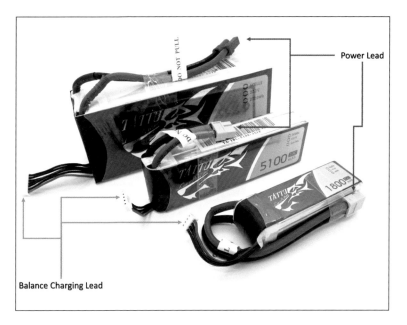

ABOVE LiPo batteries come in a variety of shapes and sizes to suit any application. *(Author)*

What the numbers mean

When you're looking for a battery there are various specifications that define its parameters.

Cell count

The main classification of batteries is the voltage. In lithium-polymer batteries this is defined by the cell count. Each cell in a LiPo battery has 3.7V and is made of a soft, thin, rectangular casing that contains the battery chemicals. Due to the nature of these chemicals the maximum voltage of a fully charged cell is 4.2V and the minimum cell voltage should never drop below 3V. If the voltage goes beyond this range it can cause permanent damage to the battery, or in the worst situations it could become unstable and explode. For this reason special balance chargers are used to ensure the battery is charged in a safe manner.

To get higher voltage out of LiPo batteries the number of cells inside is increased, with the most common arrangement being a three-cell battery, which has a voltage of 11.1V. The cell count of batteries is written as 1S for a one-cell battery, 2S for a two-cell battery, and so on. For drones battery requirements typically range from 1S (used on micro drones) to 6S (usually used on heavy-lift drone platforms). Using a higher voltage (or cell count) means that you can get more power out of your motors. However, some motors and ESCs have a specific range of voltages that they support, so you must make sure not to use a battery with a voltage that's too high – otherwise you might burn out your motor or ESC.

Cell count	Nominal voltage	Minimum voltage	Maximum voltage
1S	3.7V	3.0V	4.2V
2S	7.4V	6.0V	8.4V
3S	11.1V	9.0V	12.6V
4S	14.8V	12.0V	16.8V
5S	18.5V	15.0V	21.0V
6S	22.2V	18.0V	25.2V

Sometimes you'll see the cell count defined by a number like '3S2P', in which the first number (before the S) defines the number of cells connected in series (3S). The second number defines the configuration of how many cells are connected in parallel (2P). In order to get more capacity and higher discharge rates some batteries have more cells connected in parallel, which increase the capacity and discharge rate while keeping voltage constant. In the example of a 3S2P battery, the battery actually has six individual cells.

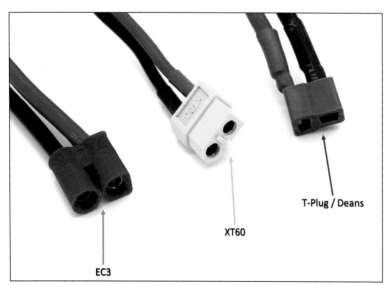

BELOW The most common types of battery connectors used with LiPo batteries. *(Author)*

Milli-Ampere hours
The amount of energy or the capacity of a battery is defined by its mAh (milli-Ampere hours). Depending on how long you want to fly you'll need to have more capacity in your battery. However, the more capacity you add to the battery the heavier it gets and the harder the motors will need to work to lift the extra weight, which will in turn drain the battery faster. So for a given aircraft there's usually an optimum battery capacity that you should use to get the best flight times. Further information on how to choose a battery will be found in the 'Choosing components for your drone' section.

C rating (discharge rate)
The C rating for a battery represents its discharge rate, or in other words the maximum amount of energy you can get out of the battery at a given time. To get the maximum current output figure in Amperes you simply multiply the capacity in Ah by the C rating – hence a 2,200mAh (2.2Ah) battery rated at 25C will have a maximum discharge rate of 55A (2.2 x 25). So when you're choosing a battery you need to make sure that it's able to output sufficient current to power all of your motors.

If you plan to fly an acrobatic or racing drone it's better to have higher C rating batteries, as this means you can send power to the motors faster and use much more powerful motors. However, the higher the C rating for a battery the heavier it will get. Consequently drones used for aerial filming often use lighter batteries with a lower C rating so that they can fly for longer.

When looking at battery C ratings you'll often find a 'peak' or 'burst' C rating that's higher than the continuous C rating. This peak or burst value is what the battery can handle for short periods (typically less than 30 seconds), and shouldn't be used when calculating if your motors will draw too much current.

Battery connectors
There are several standards when it comes to connecting the battery to other gear. Some manufacturers prefer to use a certain style of connector over another, so sometimes when it comes to buying a battery try to make sure it has the same connector as the other R/C equipment to which you'll connect it. That

way you don't need to solder on a new one each time you get a new battery. However, if a battery doesn't have the same connector as other equipment it's still possible to solder another one on to it. The most popular connectors used in the drone hobby are XT60 and T-Plug. You'll always find female connectors on batteries.

LiPo battery safety and care

All batteries have a lifespan in terms of the number of charge and discharge cycles. In order to get the most life out of LiPo batteries you should follow the guidelines regarding charging and storing them. It's also important to note that due to the higher energy density of LiPo batteries compared to others they're slightly more volatile, and there's a slight chance of them becoming unstable and possibly even starting a fire if you don't look after them. However, don't be put off by the warnings. If you take care of them properly during storage and charging, and make sure you buy brand name batteries that have undergone relevant certification testing, you've nothing to worry about. Most of the stories you read on the Internet about exploding batteries involve either inferior battery brands or improper care or usage.

As your battery gets old it might not last as long as it used to between charges and this will be evident during your flights, as you'll need to land sooner. However, later stage warnings occur when the chemistry inside the LiPo batteries becomes unstable and causes the cells to expand or puff up. If you encounter a battery that shows signs of this you should stop using it immediately and dispose of it in a *safe* manner. You can't simply put your LiPo batteries in the rubbish – you need to contact your local recycling centre to see if they accept lithium-polymer batteries and what other facilities are available.

Storing LiPo batteries

When you aren't using your batteries for an extended period there are some things you should consider. The first is to never store LiPo batteries fully charged or fully discharged, as this can upset the chemistry inside the battery and end up damaging it. Your batteries should be stored at 50–70% charge, or as close to

ABOVE You should always store and transport your LiPo batteries in 'LiPo Guard' battery bags. *(Author)*

3.85V per cell as possible. Many of the more advanced battery chargers include a storage function that ensures each cell in the battery is at the correct voltage for storage. However, if you don't have this facility you can either connect the battery to your drone and run the motors until it's at approximately 50%–70% charge, or you can use a battery voltage monitor to get each cell to approximately 3.85V.

You should also keep your LiPo batteries in a LiPo-safe bag for added security. These bags are designed to be fireproof (or fire-resistant, depending on brand) to prevent fire from spreading if something untoward were to happen.

When storing your batteries make sure that they're kept in a cool, dry place. In the UK this is fairly easy, as it never gets too hot. Also, don't leave a LiPo battery in the boot of your car on a hot sunny day, since exposure to heat for long periods can damage its chemistry. If you aren't going to be using your LiPo batteries for a long time it's also recommended that you check their voltages every month or so, just to ensure they're all within 50%–70% of maximum charge, as you might need to perform a top-up charge from time to time.

Charging LiPo batteries

Due to the design of LiPo batteries and their number of cells you need to use a special LiPo balance battery charger. Ideally it would be great to charge your battery as fast as possible so that you can get your drone back into the air, but charging batteries too fast can damage them, or cause them to become unstable and

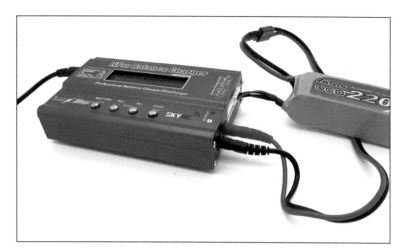

ABOVE LiPo batteries are charged using special LiPo chargers. The charger constantly monitors the voltages of each LiPo cell via the balance lead connector, and charges them all to ensure that every cell has the same voltage. *(Author)*

possibly even explode, so all batteries specify a maximum charge rate. Most modern LiPo batteries have a maximum charge rate of 5C.

To maximise the lifespan of your battery it's best to charge it as slowly as possible. Many people suggest that 1C is the maximum rate you should use, though this is really only true for some low-grade LiPos. Some of the more premium LiPo brands, which have passed stringent battery safety tests, are capable of being charged at higher C rates (typically up to 5C) with no adverse effects on battery life or performance. However, please read your battery's documentation to check its specific maximum charge rate.

It's important to never leave a charging LiPo battery unattended. Although it's very rare the battery or charger could malfunction, in which case monitoring the charge process allows you to step in and unplug the charger to prevent a catastrophe. During charging you should regularly check the temperature of both charger and battery, and if either get very hot to the touch you should stop charging immediately and consult an expert (such as your local hobby shop). Some advanced chargers include certain safety features built in, and are able to monitor the temperature and react accordingly. I'd therefore suggest you invest in a good quality charger, as it's a piece of equipment that's used throughout the hobby.

Gimbals and cameras

Cameras

Most drones have some sort of camera on board, as viewing the world from above is their primary role. The choice of drone and camera are dependent on one another, as a drone only has enough power to lift a certain amount of weight. If you need to carry a specific camera you will need to choose the relevant components, such as motors, frame etc, to be able to lift it. A camera that isn't stabilised doesn't produce very compelling footage, so often you'll also have to consider what gimbal you need to ensure your footage is stable and smooth.

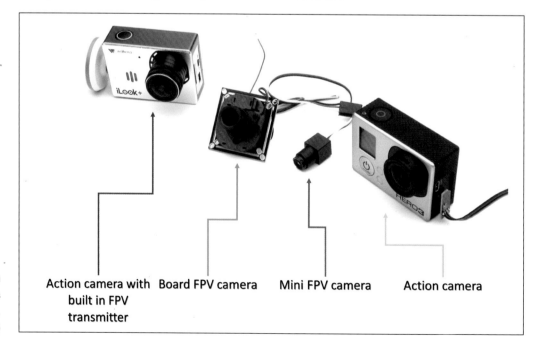

RIGHT You can use a wide range of cameras with your drone. *(Author)*

Action camera with built in FPV transmitter | Board FPV camera | Mini FPV camera | Action camera

FPV cameras

A first-person-view (FPV) camera's main purpose is to provide a real-time view from on board the drone that you can use to fly the plane remotely, as if you're sitting inside it. Typical FPV cameras are based on very small and lightweight security camera systems; their quality, however, is usually poor when compared to consumer digital cameras. Due to this poor quality FPV cameras are often used only to control your aircraft, while a second higher-quality camera (such as a GoPro) is used to record the footage in HD.

An FPV camera usually just has a few wires sticking out which are used to power it and to output the video signal that you connect to your FPV video transmitter. More details of the overall FPV system are discussed in the 'First-person view' section. Also, because such cameras are used to fly the aircraft they're mounted directly on to the airframe, with no gimbal for stabilisation, as this allows the pilot to get a feel for how the aircraft is reacting as it's bumped by the wind.

Some pilots (particularly of fixed-wing aircraft) like to put their FPV cameras on a pan-tilt mount, which provides them with the ability to look around while flying, adding an extra level of immersion to the flight. And if that's not enough you can even add head-tracking so that the motion of the camera is directly connected to your head movements, so that when you turn your head the camera on the aircraft does the same thing, which is really cool.

For those pilots looking for total immersion it's also possible to integrate 3D FPV cameras on your drone. Three-dimensional FPV cameras are actually two cameras positioned next to each other, just like our eyes. By overlapping the images from the two cameras and projecting it to your FPV headset it gives the illusion of 3D. This enhances the flying experience even further.

FPV cameras have two types of imaging sensors: CCD (charged coupled device) and CMOS (complementary metal oxide semiconductor). CMOS cameras are more expensive and heavier than CCD but are typically better suited for FPV applications due to their large dynamic range and greater light sensitivity. This is important when flying, as with some poorer quality cameras the ground will be very dark and the sky very bright, making it hard to see anything on the ground – including obstacles you might fly into. CCD cameras are getting better with time, and with improvements in the processing software some can achieve the same performance level as CMOS cameras, while costing less and being lighter.

The main specification to worry about is the resolution of the camera, as this gives an indication of the clarity of the picture. Often quoted as TVL (TV line), you typically want to

BELOW A NerdCam 3D FPV camera consists of two cameras placed side by side, enabling a three-dimensional FPV experience. *(Rolando Cerna)*

ensure your camera has a resolution of at least 600TVL. There are some newer FPV cameras with resolutions as high as 1,200TVL, which work well with high-resolution monitors or FPV goggles.

Action cameras

Action cameras such as the GoPro series are very popular, because they allow you to record high-quality videos (up to 4K on some) with a small, lightweight camera. The videos are recorded on to an SD card in the camera. GoPro cameras are popular for use on drones as you can connect them directly to your video transmitter, allowing you to view what the camera is seeing in real time and thereby making it much easier to know exactly what you're recording. You'll find that most gimbals are designed for the GoPro range of cameras. However, there's also a range of 'clone' GoPro-type cameras designed by other manufacturers that use the same form factor so will also fit into most GoPro gimbals. Depending on your budget and quality requirements it might therefore be better to go for a cheaper camera to start with, as they're often good enough for most applications. Note that a recent GoPro firmware upgrade has disabled FPV Video Out for some models, so it's worth checking before buying one.

There's a range of smaller action cameras that's also popular, usually seen on smaller drones such as FPV racing quadcopters. These cameras, widely known as 'keychain' cameras, are remarkably small and lightweight (between 20–40g) and are therefore ideal to mount on just about any size of drone. Although some of them are able to record in full HD that provides great quality, they're still not as good when compared to larger cameras with bigger lenses and sensors.

Cameras for photos

The cameras mentioned above are great for shooting videos. However, for other more professional applications, such as mapping or aerial photography, small digital cameras or DSLR (digital single-lens reflex) cameras are often used, as they're able to take much better quality photos. There's a huge range to choose from so we won't go into any details, but their most important aspect after image quality is their control interface, as you need to be able to remotely trigger the camera so that it can take a photo while flying above you.

Camera triggering

As your camera is mounted on your drone it's obviously not possible to press the shutter button to take a shot. Consequently a remote camera trigger is used. There's a variety of camera-triggering mechanisms but most of them are manufacturer-specific, so will only work with a certain brand of camera, such as Sony, Nikon or Canon. Before buying one check the description, which will often tell you if it's compatible with your camera or not. There are four primary types of camera-triggering device:

Infrared – This method uses an infrared LED (IR LED) that flashes in a particular sequence and at a particular speed that's detected by the infrared receiver port on your camera. This is a common method of taking a photo by drone, as most of the better cameras include an infrared port for remote triggering. You'll often receive an infrared LED handheld remote when you buy a camera, but this is different from the one you'll use with your drone. The IR LED is connected to your autopilot or R/C receiver and will trigger when you or your autopilot tell it to. If using an IR LED you'll need to mount it in view of your camera's IR receiver port.

Direct cable – As some cameras don't have an IR port, or it's difficult to mount an IR LED, another option is to use a direct cable connection. This method is most popular with

BELOW Example of an infrared camera trigger device mounted on a camera with some double-sided tape. The trigger is controlled by the autopilot or R/C receiver. *(Author)*

Canon cameras loaded with CHDK firmware (Canon hacking development kit), which allows the camera to be triggered via a USB cable connected to your autopilot. Many Sony cameras can also be triggered with the use of a multi-USB or LANC cable.

Mechanical – This involves directly mounting a servo on to your camera to physically depress the camera shutter button, and is something that you should only resort to if the other methods don't work with your camera.

Timer based – Some cameras have the ability to take photos repeatedly at set intervals, so you can start taking photos by pressing the button when your drone is on the ground and it'll continue to take photos at set intervals as it's flying. This method is often not ideal, as it means you often have a bunch of extra photos that you have to sort through to find the ones you want, but it's still a way of capturing aerial photos if the other methods don't work.

If you plan to use your drone for mapping it's highly recommended that you use a camera that supports a direct cable or IR camera shutter that can be controlled by your autopilot. The reason for this is in order to increase the value of your generated images, as some autopilot systems have the ability to record the GPS location of where the photo was taken. This is called geo-referencing. Using geo-referenced images usually results in better quality results for mapping or creating 3D-terrain elevation models.

Vibrations

Vibrations can compromise the quality of the images captured by your drone, resulting in blurry photos and wobbling-jelly-effect videos. So when mounting your camera it's important to make sure you isolate it from vibrations by using vibration-damping mounts or soft double-sided tape. The good news is that most of the gimbals on the market today already include proper vibration-damping features to ensure your videos and photos are free from such problems. Another way to minimise vibrations is to make sure your propellers are properly balanced, as discussed in the 'Propellers' section.

Gimbals

A gimbal is a device that keeps your camera stabilised while your drone is moving about in the wind. Traditionally all gimbals were servo-based; however, in recent years brushless gimbals have become the de facto type to use with drones, as they have superior stabilisation performance. Many of today's brushless gimbals will keep your camera stable to within 0.1° of error as the drone is flying around.

Brushless gimbals make use of brushless motors that are essentially the same as those used to drive the propellers, the main difference being that the stator-winding sequence of the gimbal motors is done in a specific way to have a much lower KV rating, to maximise torque. When referring to gimbals they're defined by the number of axes, most being either two-axis

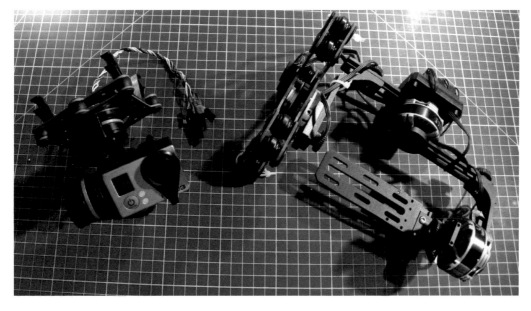

LEFT Most gimbals you'll find today use brushless motors, to keep the camera level as you're flying. *(Author)*

or three-axis. This refers to how many axes the gimbal can stabilise – a two-axis gimbal can stabilise the roll and pitch axes and will have two motors, but a three-axis gimbal will stabilise roll, pitch and yaw and have at least three motors.

Gimbal weight limits

Because the brushless motors used on gimbals have maximum torque limits a given gimbal will only cope with a certain range of supported camera weights. For example, some gimbals use specific gimbal motors that will only support cameras weighing around 500g, while others use smaller motors that will only support cameras to a weight of 300g. So even if you can fit your 400g camera on a GoPro-size gimbal it's more than likely that it won't work, since the motors won't be powerful enough to keep the camera stable. Consequently when purchasing a gimbal you should always check that it'll be able to support your specific camera's weight and dimensions.

Balancing your gimbal

In order to get the best performance out of a brushless gimbal the most important set-up step is to properly balance it *with your camera mounted*. This involves adjusting the position of your camera slightly so that when you tilt or roll the gimbal it keeps its position and doesn't fall to one side. This balancing process often isn't required with some camera-specific ready-to-use gimbals, as the mounting location is pre-fixed to the correct position so that the gimbal is already balanced.

With some smaller, lighter cameras the motors are often powerful enough to overcome slight balance issues. However, when using heavier cameras, such as a DSLR, the balance needs to be almost perfect in order for the gimbal to work properly. Often things like changing the lens on your camera will require you to rebalance the gimbal system. When the gimbal is perfectly balanced your camera won't move no matter what angle you move it to on the gimbal when the gimbal isn't turned on, as the camera's centre of gravity is in the exact centre of the gimbal.

1 To balance your gimbal, you should start with the roll axis. With the camera mounted on the gimbal and the gimbal turned off, see if your camera rolls to the left or right. If it rolls to the right, move it further left along the camera mount, and if it tilts to the left move it rightwards.

2 Once you think the camera is balanced you can double-check by moving it to a level, horizontal position. If it stays there when you let go then your balance is good enough.

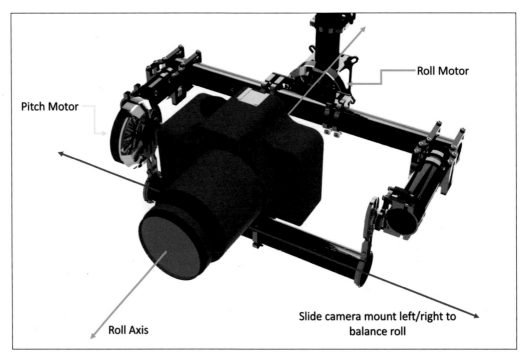

RIGHT To balance the roll axis you need to move the camera left or right to ensure its centre of gravity is aligned with it. *(Author)*

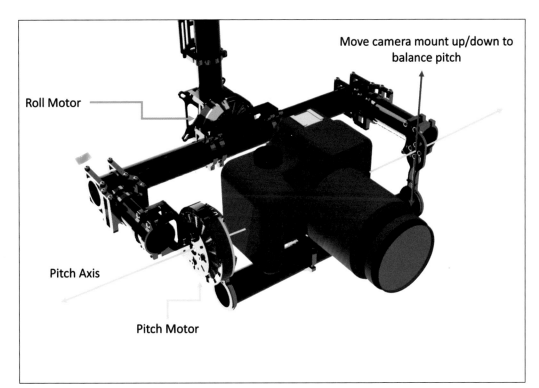

LEFT The pitch balance of the gimbal is adjusted by moving the camera up and down to align its centre of gravity with the yaw axis. *(Author)*

3 The next step is to balance the pitch motor. This is done by moving the camera-mounting platform up or down. Hold the camera so that it's facing forwards, and when you let it go it should stay there if the gimbal is balanced. If it pitches forward move the camera upwards, if it tilts upwards move it down, and so on.

4 Most gimbals also have the ability to move your camera forwards and backwards, to aid in getting it balanced. Start by placing your camera pointing downwards. If it rotates forwards or backwards you'll need to slide it backwards or forwards accordingly on the camera platform. You shouldn't proceed to balance the yaw axis until the roll and pitch balance is perfect.

5 To balance the yaw axis you'll need to slide the yaw motor forwards or backwards. Position the gimbal so that the camera is looking forwards and you're standing behind it looking in the same direction. First tilt the entire gimbal downwards. If the yaw axis starts to rotate then there's too much weight at the back of the gimbal and you need to slide the gimbal yaw axis forwards.

6 Similarly, if you tilt the entire gimbal backwards and the gimbal rotates then there's too much weight in the front, so you need to slide the gimbal yaw axis backwards. Once the gimbal yaw axis is balanced it shouldn't move when you tilt the entire gimbal forwards, backwards, left or right.

BELOW Balancing the yaw axis of the gimbal involves sliding the yaw motor forwards and backwards until this axis is aligned with the gimbal's centre of gravity. *(Author)*

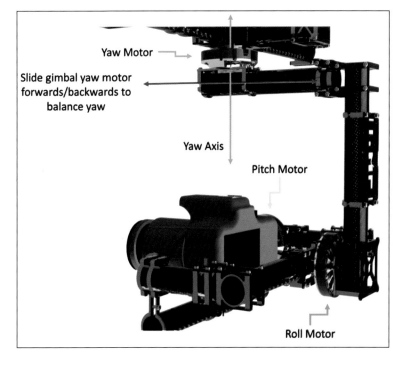

Gimbal controller
Brushless gimbals need a brushless gimbal controller, which sends the specific signals to control the motors precisely. Most gimbal controllers make use of an inertial measurement unit (IMU) that measures the angle of the camera at over 500 times a second and reacts to keep the camera level. This IMU is essentially the same as that used on your flight controller (and what's inside your smartphone), and makes use of accelerometers and gyroscopes.

Some gimbal controllers use two IMU sensors to get even better stabilisation results. In this case one IMU is mounted on the camera platform to measure the camera angle and the other is mounted to the gimbal base (or drone), which allows the controller to read data from two sources in order to provide better levels of stabilisation.

Other gimbal controllers make use of precision potentiometers, which are rotary position sensors that measure the camera angle to keep it level. The advantage of using potentiometers is that they don't require any sort of calibration to initialise them during gimbal start-up. However, they're often more complex to integrate into the gimbal and are only really utilised in some ready-to-use gimbals.

First-person view

Using a small camera, a wireless video feed and either a monitor or video goggles on the ground, first-person view (FPV) puts you in the virtual pilot's seat of your remote-control aircraft. FPV flying has become very popular in recent years, as it gives you the sensation of actually being in the aircraft – especially if you're wearing video goggles.

The other main use of FPV is if you're creating videos, since having a wireless video feed from your camera allows you to see what you're recording, making it much easier to get the perfect shot. For film work using a video screen is much more common than video goggles, as in professional film work there are usually two drone operators – a pilot who controls the drone and a gimbal operator who operates the camera; using a monitor allows both to see what the camera's recording and where the drone is.

In terms of wireless video transmission this can be done via analogue video transmitters, which are more popular for hobby use, and digital video transmission, which is more costly but produces better quality imagery. Analogue video systems are by far the most popular due to their size and low cost. Digital video transmitters and receivers are currently very expensive in comparison, but their ability to generate real-time high-definition video makes digital FPV the preferred system for some individuals, particularly in professional applications such as film work.

Digital video systems for airborne use have only become small enough to be practical in recent years and remain very expensive, with a typical transmitter and receiver costing about £1k. In the future, however, as with most new technologies we should see the price and size of digital video systems reduce to a level that makes them more competitive with hobby-level analogue. By contrast an analogue video transmitter receiver system can be purchased today for under £30, which explains why it remains so much more popular.

A basic FPV set-up consists of a camera connected to a video transmitter that wirelessly transmits the signal to a receiver back on the ground. An overview of how everything is connected together on a basic FPV system is shown opposite.

Nowadays most FPV equipment can support both PAL and NTSC video standards, so this is less of an issue than it used to be, but it's still something worth checking before you buy any FPV equipment, to ensure that everything will talk on the same video standard. So your camera and your video monitor/goggles all need to be on the same video standard (ie both PAL or both NTSC). If your camera uses PAL and your monitor uses NTSC you won't be able to get a proper video signal. PAL has a higher resolution (720 x 576 at 25 frames per second), but NTSC allows for a higher frame-rate (720 x 480 at 30 frames per second), so you can use either one depending on what you prefer.

FPV transmitters
There's a huge selection of video transmitters and receivers available on the market today, so it's important to understand some of their features in order to choose the best one for

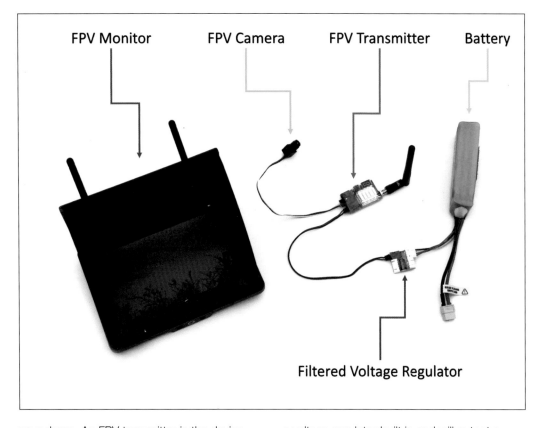

LEFT An example of the components included in a basic first-person-view set-up. The FPV camera and transmitter are mounted on the aircraft and are often powered by the flight battery via a filtered voltage regulator. On the ground you'll be able to view the live video from the camera in real time via an FPV monitor. *(Author)*

BELOW FPV video transmitters are available in many different sizes and output strengths. The one on the left requires a built-in fan to keep it cool during operation, as its high output means it generates a lot of heat. *(Author)*

your drone. An FPV transmitter is the device that transmits the images from your on-board camera wirelessly over the airwaves back to the ground, where your FPV receiver will read the signals and display the image on your monitor. .

Connecting a camera to your FPV transmitter

Although many FPV transmitters might look quite different, or have totally different-looking connectors, they'll often have all the same functions. For example, all FPV transmitters have a video input connector and a power connector to operate the FPV transmitter, but they might be located in different places. So before you connect anything you must make sure to check the documentation of your FPV transmitter and camera to make sure you know what each wire and connector is for.

Some video transmitters might include extra connectors, such as power output to your FPV camera, which is very useful. However, a side note on this is to check the specifications of the FPV transmitter power output. Some transmitters will output the same voltage as the input voltage (or the same voltage as the battery connected to them), but others will have a voltage regulator built in and will output a constant 5V. So be sure to check if the output voltage will be compatible with the voltage range of your camera to ensure the voltage won't be too high or too low.

Another aspect when trying to pair an FPV

transmitter to a camera is to look at the types of connectors. I'd suggest you try to make sure they have compatible connectors to make connecting them together as easy as possible. However, if you can't find any suitable connectors it's always possible to solder some wires together to make your own, as most FPV transmitters and cameras include some extra cables for you to do this.

An important safety note with all FPV video transmitters is to never turn them on without an antenna connected, as this can damage the video transmitter module.

Frequency channels

All aerial FPV gear in the UK, and most of Europe, runs on the 5.8GHz frequency band due to various regulations. This consists of frequencies in the 5.725GHz–5.875GHz range, which is broken down into small sub-sections of frequencies called frequency bands. Each of these frequency bands can be further divided into channels, and each channel corresponds to a specific frequency. An example of typical 5.8GHz FPV transmitter frequency bands and channels is shown in the table below:

You should also note that Band E in this table actually exceeds the ISM (industrial, scientific and medical) 5.8GHz band, and in the UK it's illegal to transmit on this frequency unless you receive special permission. However, since the table is taken from an FPV receiver it's presumably acceptable to have that functionality, you just mustn't transmit signals outside the 5.8GHz ISM range. Most CE-certified 5.8GHz FPV transmitters will only transmit on Bands A, B and F, as those bands all fall within the 5.8GHz ISM band at the legal power level (more on power levels below).

In most of this book we've assumed FPVs to be on the 5.8GHz frequency band, as this is by far the most popular. However, some pilots might want to use 2.4GHz for their FPV as it's also legal to do so in the UK, though many people prefer not to as most R/C transmitters use this frequency, which can cause interference.

Power levels

In the UK and most of Europe the maximum power level you can use on the 5.8GHz frequency band is 25mW. Power levels above that require a special radio licence or permission to use them. With the correct antennae on your transmitter and receiver you can easily get a range of over 500m, and with high-gain tracking antennae and the right conditions some FPV pilots have even managed to reach distances of over 5km using standard 25mW video transmitters. However, it should be noted that flying beyond visual line of sight is unlawful without permission in the UK. If you want to make sure you're not breaking any laws it's best to only buy FPV transmitters with a CE certification mark. If you're using a 2.4GHz FPV transmitter the maximum power level you're allowed to use is 10mW.

Increasing the power levels of your FPV equipment doesn't necessarily give you longer range, as radio equipment range follows what's known as the inverse square law. This means that if you double the transmitter power you won't double the distance of your FPV gear; rather you'll only get 25% extra. So using the correct antennae is the best way to get more range and still stay within the regulations. More details on FPV antennae will be found later in this section.

Matching an FPV transmitter to a receiver

The frequency table might look confusing at first glance but the main thing to bear in mind when choosing your components and setting up your system is that your FPV transmitter and receiver are both set to exactly the same frequency. For example, if I set my transmitter to run on Band B, Channel 5 (5.809GHz), I must set my FPV receiver to exactly that channel. You might still be able to pick up the video signal if your

ISM 5.8GHz								
	Channel 1	Channel 2	Channel 3	Channel 4	Channel 5	Channel 6	Channel 7	Channel 8
Band A	5.865GHz	5.845GHz	5.825GHz	5.805GHz	5.785GHz	5.765GHz	5.745GHz	5.725GHz
Band B	5.733GHz	5.752GHz	5.771GHz	5.790GHz	5.809GHz	5.828GHz	5.847GHz	5.866GHz
Band E	5.705GHz	5.685GHz	5.665GHz	5.645GHz	5.885GHz	5.905GHz	5.925GHz	5.945GHz
Band F	5.740GHz	5.760GHz	5.780GHz	5.800GHz	5.820GHz	5.840GHz	5.860GHz	5.880GHz

ABOVE An assortment of aircraft at an FPV meet-up. *(Patrick McKay)*

receiver is tuned to a similar frequency such as Band F, Channel 5 (5.820GHz), but the signal won't be very clear or reliable. So it's best it make sure both frequencies are set to exactly the same value.

Also, when choosing to buy an FPV transmitter and receiver (or FPV monitor or goggles with a built-in receiver) it's important to make sure they can both work on exactly the same frequency channels, otherwise they might not be compatible. Information about the supported frequency channels is often included on the description page of FPV transmitter and receiver handbooks, but if you're not sure it's best to ask at the shop before making your choice.

Flying with your friends

Because FPV transmitters and receivers are able to switch channels, this means that it's possible to have more than one FPV aircraft flying in the same area so long as they're all set to different frequencies. To ensure the least possible interference it's best to make sure the frequencies are separated as widely as possible.

When flying with other FPV pilots make sure to check with them what frequencies they're using before turning on your aircraft or FPV transmitter. If you have the same transmitter frequency set as someone else who's flying your signal will interfere with theirs and could cause a crash.

Also, if you're not flying you can always tune to the same frequency on your receiver as other pilots who are flying, to get a feed from their cameras, which is great if you just want to watch. Because FPV transmitters work in a similar way to regular TV you can essentially have an unlimited number of receivers connected to a single video feed assuming they're in range.

Changing between channels and bands

With most FPV transmitters and receivers there are two main methods for changing the channels. The first is via DIP (dual in-line package) switches, and the second is via a push-button. DIP switches are tiny switches located on your transmitter and receiver, which depending

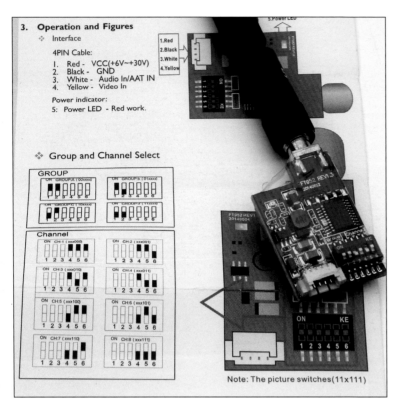

ABOVE The DIP switch channel table is included with the instructions for your FPV equipment and tells you the order in which to set the switches to select the various channels.
(Author)

on the order that they're set to will define the frequency at which the equipment will work it. To set the channels you'll need to look at the documentation, which will have a table with switch positions that you can use as a reference to set specific channels. Often you'll need to use tweezers or a small screwdriver to change the switch positions, as they're very small.

Other FPV equipment simply uses a push-button and a graphical display to show you what channel you've selected. Using a push-button is much easier as you can just use your finger to change between the channels. Again, you'll need to consult the channel table to know what frequency corresponds to each channel.

Some of the more advanced FPV receivers have the ability to quickly scan across the 5.8GHz range and automatically tune in to the channel of your transmitter, making things rather easy. This works just the same way as a TV that has an auto-tune function.

FPV receiver

As its name suggests, the FPV video receiver is the device that reads the video signals from the transmitter and outputs a signal that can be read by your video monitor or FPV headset. Standalone receivers are becoming less popular these days, as most new FPV equipment such as monitors and goggles often includes a built-in receiver, which makes setting everything up much more convenient. However, the advantage of using a dedicated FPV receiver is that you can often mount it on to a tracking antenna stand, a tall tripod or mast to ensure you get the best reception, which isn't possible if your FPV monitor has a built-in receiver. Even though many new FPV monitors and goggles include a built-in receiver they also retain an external video input connector, so you still have the option of using a dedicated receiver if you require.

DVR

Some FPV receiver equipment includes DVR (Digital Video Recorder) functionality, which lets you record live FPV footage to an SD card. But the usefulness of this is questionable, as the quality of FPV feeds often isn't great due to low resolution, and because you fly further away you might get some static. Most of the time you'll record the HD video footage free from inference with a secondary HD camera such as a GoPro or Mobius action camera, and just use the live video feed to help you fly. (Some cameras such as the GoPro have the ability to record HD video wile outputting the live video feed at a lower quality, so you can still use a single camera.)

Monitors or goggles?

The choice between using an FPV monitor or FPV goggles boils down to user preference. However, there are some situations where one display device will be better than the other.

FPV goggles are better if you plan to fly FPV for its sense of immersion, as the goggles fill your entire vision with what your on-board camera can see. FPV goggles include small screens built into a pair of goggles that places them just in front of your eyes. This gives the effect of looking at a larger screen. However, if you wear glasses using FPV goggles can be a pain, because you can't usually wear them over your ordinary glasses. Some FPV goggles include lens adapters than you can insert so that you don't need to wear glasses, but as soon as you take them off you won't be able to see anything in the event of your FPV signal cutting out. There are, however, some headsets that have sufficient space for you to

RIGHT Wearing video goggles for FPV piloting is a totally immersive experience. *(Author)*

wear glasses in conjunction with them, which solves these problems. The other alternative, of course, is to use contact lenses.

When using FPV goggles you might find it disorientating at first, but once you get used to them they really do get close to giving you the exhilaration of actually flying!

Using an FPV monitor is usually better if you're using the video feed more as an auxiliary function on your drone, which is commonly used for aerial photography or filming applications. This lets you see what the camera is capturing but still gives you the option of being able to look up to see where your drone is flying, allowing you much better situational awareness. Many FPV monitors also include a built-in or optional sunshield, which makes it much easier to see what's on the screen in direct sunlight, so it's always worth checking this before purchasing a monitor to make sure it either includes one or has the option of adding one.

BELOW Using an FPV monitor while flying gives you the ability to see what your camera is recording while still being able to look up at your drone to maintain good situational awareness. *(Author)*

RIGHT **The diversity video receiver built into the FPV monitor allows you to connect two antennae to get a better video signal.** *(Creative Commons License)*

Diversity

Some FPV receivers have a feature called 'diversity', which means that it allows you to use more than one antenna, and the receiver will automatically switch to the antenna that's getting the best reception. Diversity is a great feature that results in your receiver getting a better video signal.

Antennae

All the FPV equipment you purchase will come with a basic antenna, but many FPV pilots choose to replace these with antennae better suited to FPV, or that provide greater range. There are consequently several concepts that you need to understand to help you choose the ideal FPV antenna for your drone, which are discussed here.

In order to understand the main concept it helps to think about the available energy in your FPV signal as being like a water balloon. Depending on the antenna that you use, it will change the shape of this balloon but the amount of energy (or water in it) will remain constant.

Directional antennae

Still keeping in mind the idea of a water balloon, if we stretch it out in a specific direction it will cover a greater distance but become thinner. This is the concept that directional antennae are designed around. They provide you with much greater range in the direction they're pointing, but if you fly to either side of the beam you won't pick up any signals. So directional antennae sacrifice width for extra range. Although these antennae provide great range you need to ensure that they always point at your aircraft, so you'll often find directional antennae mounted on antenna-tracking stations so that the antenna always points at your aircraft as you're flying it.

Omnidirectional antennae

In contrast to directional antennae, omnidirectional antennae have a lower range but work in a full spherical pattern; so you can think of this as a water balloon resting on the ground, where it spreads evenly in a circular manner but doesn't extend particularly far in any direction. Omnidirectional antennae are often the preferred choice, as you don't need to care about what direction your antenna is facing.

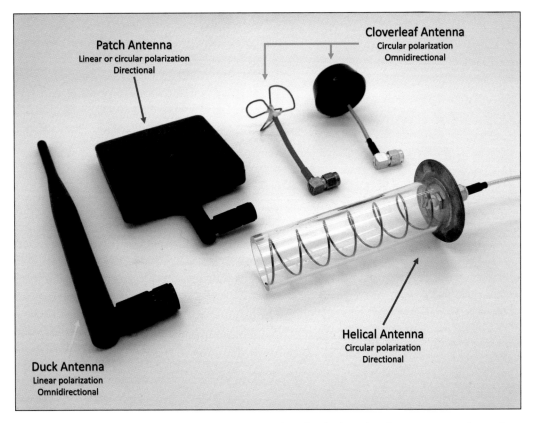

LEFT There are several types of antenna you can use with your FPV gear, each having its own characteristics. *(Author)*

BELOW Illustration showing a linear polarised wave. Although this diagram might seem confusing the only part you need to care about is the red wave, which is the signal that's created on a single linear plane. The blue and green waves are two components used to create the polarised wave. *(Wikimedia Commons)*

Diversity

Although this is actually a feature of your FPV receiver, diversity means you can use more than one antenna on your receiver and it will automatically switch to the one that's receiving the best signal. This allows the best of both words, as you can have one directional antenna to get long range, and an omnidirectional antenna to get more coverage when you're flying closer.

Linear polarisation

When an antenna uses linear polarisation it means that the radio waves are created on linear planes. Most systems, such as your R/C transmitter, use these types of antennae and they work well. Because the signals move along these linear planes, the best signal is received when the planes are aligned. This occurs when the antennae on both the transmitter and receiver are aligned (*ie* both facing upwards). As your aircraft is flying it doesn't always remain exactly level, so when you turn the antennae are no longer aligned, which can cause a separation of the signal. For this reason many FPV pilots prefer to use circular polarised antennae as discussed below. However, depending on how you're flying multirotors generally stay more level than fixed-wing aircraft, so you can often get a perfectly adequate signal by using the standard linear polarised antennae that are frequently included with FPV transmitters.

Circular polarisation

Circular polarised antennae have a slightly different radiation pattern that's circular. Their

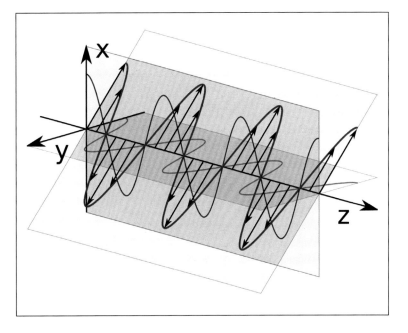

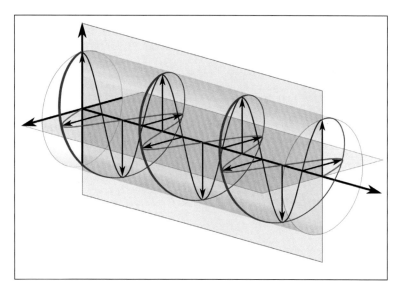

ABOVE Illustration of the wave emitted from the right-hand circular polarised (RHCP) antenna. The blue and green sections are the horizontal and vertical components of the circular polarised wave (red circular region) emitted from the antenna. *(Wikimedia Commons)*

turning and very rarely being exactly level they guarantee a reliable signal even when banking at high angles. Furthermore, circular polarised antennae have better multipath rejection. Multipath is when a signal bounces off an object such as a building and the same part of the signal arrives twice, manifesting itself as ghosting in the video, where the older video signal is overlaid on the current signal. The main disadvantage of circular polarised antennae is that their range is lower than that of linear polarised antennae that are aligned.

When buying circular polarised antennae it's important to make sure that the antennae on your transmitter and receiver are polarised in the same direction. What I mean by this is that the circular polarised antennae can form the wave as clockwise propagation (left-hand circular polarised, or LHCP) or anticlockwise propagation (right-hand circular polarised, or RHCP). You need to ensure that both antennae have the same polarisation or else they won't be able to communicate. Details of the direction of polarisation will usually be specified on the product documentation, but if not you can just inspect the antenna lobes. Looking at the antenna from the side you'll be able to tell which it is, as RHCP will have the antenna wire tilted to the right. If you aren't sure when buying

main advantage is that the signal is less reliant on the antennae on both transmitter and receiver being parallel, as the signal emitted from the antenna is circular and not on a single plane as with linear polarisation.

Circular polarisation antennae are ideal for most FPV pilots, since despite their aircraft

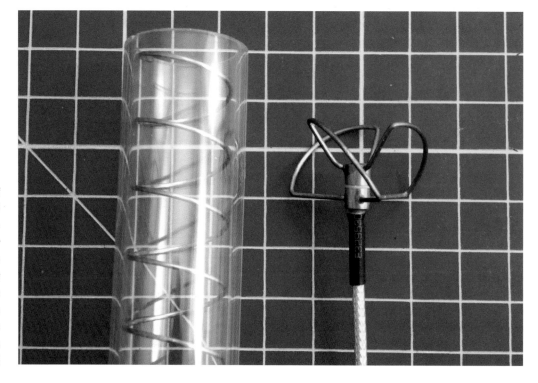

RIGHT Example of helical (right) and cloverleaf (left) left-handed circular polarised antennae (LHCP). You can visually identify the direction of circular polarisation by inspecting how the antenna lobes are angled. *(Author)*

an antenna you can always ask at the shop or check the product description.

RP-SMA and SMA connectors

When it comes to buying antennae for FPV transmitters or receivers there's much confusion between SMA (sub-miniature version A) and RP-SMA (reverse polarity sub-miniature version A) connectors, as they look very similar. Individual manufacturers generally tend to prefer either SMA or RP-SMA connectors on their equipment, so when choosing an antenna for your FPV transmitter or receiver make sure that the connectors of both are either SMA or RP-SMA, as you can't use an SMA connector with an RP-SMA one. An illustration showing the difference between their connectors is shown on the right.

Frequencies

We've already discussed frequencies in the FPV transmitter section, but in terms of antennae you'll find that they're designed to be used with specific frequencies. For instance, you won't be able to use an antenna designed for 2.4GHz equipment with a 5.8GHz video receiver, as it won't pick up any 5.8GHz signals.

Antennae designed for lower frequencies will be bigger than those designed for high frequencies. To get technical, when some manufacturers design a 5.8GHz antenna there will be a specific frequency channel within the 5.8GHz band that this antenna is designed for. So if you can set up your FPV to work on that frequency you'll be able to get the most out of your antenna and FPV gear. However, because the improvement can be negligible manufacturers don't always publish this information.

As a rule of thumb it's best to try to transmit as close to 5.800GHz as you can to get the best range, as this is the frequency the antennae are usually designed for.

Antenna gain

The gain of an antenna will give you an idea of how sensitive it is. As a warning, you should never use a high-gain antenna on your transmitter as it could cause the equivalent transmitter power to become unlawful to use. So it's best to stick to either the standard antenna included with the transmitter or to

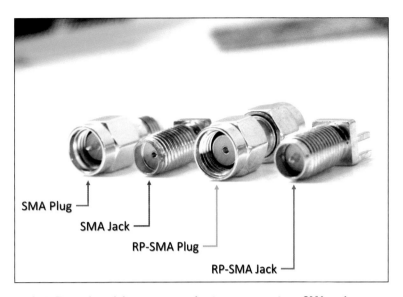

ABOVE Examples of the two types of antenna connectors, SMA and RP-SMA. Typically you'll find the jack connector on the actual device and the plug on the antenna. You can only connect an SMA jack to an SMA plug, and an RP-SMA plug to an RP-SMA jack. *(Author)*

replace it with a standard transmitter cloverleaf antenna if required. You can, however, use as high a gain as you want on your receiver as this doesn't send out any signals. So often you'll find very high-gain antennae – such as patch antennae, Yagi or helical antennae – on the receiver of long-range FPV set-ups.

Antenna placement on your aircraft

The placement of your antenna can affect the range and quality of your video signal, as some of it can be blocked by things such as the aircraft body. If using linear polarised antennae make sure that they're parallel to each other to ensure the largest possible overlap between the transmitter and receiver antenna. This is less important for circular polarised antennae as discussed previously.

The other factor to consider when mounting your transmitter antenna is to make sure that it isn't too close to objects such as motors or carbon fibre plates that could possibly block the signals being sent to the ground receiver when you fly at certain angles. Let's assume your antenna is mounted on the back of your aircraft, and you're flying away; the antenna has a nice clear view back to the home location of the receiver. But when the aircraft turns to face the home location to fly back, your aircraft will be between the transmitter antenna and the receiver, which could cause a problem. I had this situation occur once which resulted in my having to fly my multirotor back home with the front facing away from me, to maintain a video signal. Fortunately this isn't a problem with a quadcopter, but on a fixed-wing drone it isn't possible. So depending on where your FPV transmitter is mounted it might be better to have the antenna facing upwards, or preferably downwards. Just choose a position where the least equipment gets in the way when the aircraft is flying at certain orientations.

On-screen display

Often it's useful to overlay some flight information on top of the video feed. Such data can include an artificial horizon, vehicle speed, the direction to home and battery information. This is known as an on-screen display, or OSD. Most OSD devices connect to your autopilot and your video transmitter, or are built into an autopilot system, since the data that's displayed comes from the on-board sensors. You can think of an OSD as being like an HUD (heads-up display) that fighter pilots use. It's an entirely optional add-on to your FPV experience. Some pilots prefer using them while others feel they just clutter the screen.

A number of OSD units are available on the market. Some are standalone units while others are fully integrated with your flight controller. Typically the standalone units will provide limited information such as battery voltage, whereas OSD units that can communicate with your flight controller are able to display any information that your autopilot uses. When choosing an OSD module it will usually specify which flight controllers it's compatible with and provide connection diagrams and set-up instructions. Many OSD units also include some configuration software that lets you customise the information that's displayed and how it's displayed.

Choosing components for your drone

In this section we'll discuss the thought process to pursue when choosing components for your drone build. There are no rules in terms of what component you should start with, and many components are dependent on each other, so it's best just to start with something and work from here.

The first thing to consider is the application of your drone: will it be used for mapping, filming or just for fun flying? So it's important

RIGHT Example of an OSD overlay on the live video feed from on board the drone, showing key information such as altitude, aircraft speed and battery capacity remaining.
(Patrick McKay)

to consider what the drone will be required to carry, known as its payload. If you're building a drone purely to fly for fun it won't be required to carry anything other than the battery, but if you're using it for filming you might want to carry a specific camera, which will be mounted on to a gimbal to keep it stable, and this will add extra weight; so you'll then need more powerful motors. The application of your drone will also dictate what autopilot you can use. If you want to build a mapping drone an autopilot with mapping-specific features will be needed. However, if you plan to build a racing drone then buying a cheaper flight controller with only basic stabilisation features would be best.

Once you've chosen what payload you'll need to carry, and you know the estimated weight and size of everything, including batteries and gimbal etc, you can start to consider which motors and propellers will be required to lift it all. Once you have an idea of the size of the motors, ESCs and propellers required you can move on to choosing the type of battery you'll need to power everything. The final step will be to find a suitable aircraft frame that'll be strong enough and have enough space to lift all your gear.

Once you've gone through this process you might need to revisit some steps to refine your selections, as some components may prove to be too heavy, or you might need slightly bigger motors, or want to reduce the capacity of your battery to optimise your component selection. Based on your budget you might also need to refine the component selection itself.

However, this process isn't written in stone, and you can start with any aircraft frame if you already have one, or you particularly like the look of something, and work from there. Similarly, if you already have a bunch of motors you can try to build a drone around those, so the selection process set out above should just be treated as a rough guide. Further details on how to consider the choice of specific components is discussed in more detail throughout this section.

Choosing your motors, propellers and ESCs

Before you can decide on what motors to use you'll first need to have an estimate on how

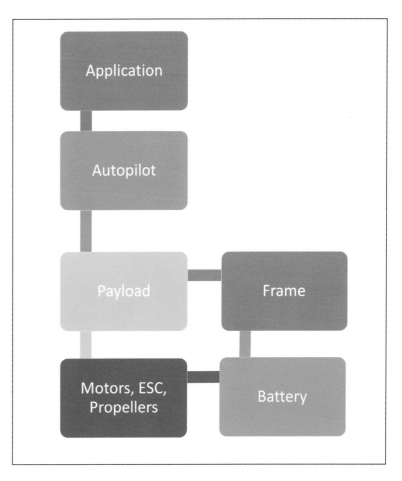

ABOVE The process of choosing parts for your drone doesn't need to follow any specific order, as most components affect one another. *(Author)*

heavy your drone will be, or what you need to carry. That way you'll know how powerful your motors need to be in order to propel the drone.

Multirotor motor and propeller selection

When choosing motors for your multirotor drone, the general rule to follow is to make sure your motors/props will produce enough thrust to be able to hover the drone at around 50–65% throttle; so if your quadcopter has a total weight of 1kg, since a quadcopter uses four motors each motor will need to produce at least 500g of thrust, to give a total thrust of 2kg (4 x 0.5kg).

The reason for this is that since multirotors use the motors' thrust to control the drone, you need to have enough thrust to ensure you can react to gusts of wind etc. In reality if you're using your drone for mapping or filming applications, where you don't need lots of manoeuvrability, you could get away with having a slightly lower amount of thrust so that you hover at around 70%, but I wouldn't suggest flying a multirotor that required more than 70% thrust to hover as it could lose control if the wind is too strong.

Once you have an estimate on the total weight of your multirotor you'll need to look through the thrust data tables (see Appendix) to find a suitable motor and propeller combination that will be able to produce enough thrust (2 x more thrust than the estimated weight of your multirotor). Since multirotors also use more than one motor you should also decide on how many motors you plan to use, so that you know how much thrust each motor will need to produce – simply divide the total thrust required by the number of motors.

Once you've found a few motors that can generate the required thrust the next step is to compare them in terms of cost and efficiency (as discussed later).

Fixed-wing motor and propeller selection
When building a fixed-wing drone the choice of components can be quite easy, as most fixed-wing planes are sold with the best motors, propellers and ESCs already included. If not they'll often provide details of the recommended equipment to use. However, we'll still discuss some general pointers for choosing a motor, prop and ESC.

If you want to choose a motor based on thrust values a very rough guide for fixed-wing drones is to select a motor/propeller combination that'll produce as much thrust as the total weight of the drone (ie a 2kg plane including payload must have a motor that produces 2kg of thrust). However, this is a very crude method and it would be better to choose a motor based on the power it produces in Watts, as this is a better measure of performance.

To choose a motor for a fixed-wing drone we need to have an estimate of the total weight of the aircraft. Based on this estimated weight we'll then choose a motor that provides sufficient power in terms of Watts. But before we do that we need to ensure that our aircraft wing will be able to lift all of our equipment. We do this by finding the wing loading, which tells us how much lift each portion of the wing must produce. The wing loading is useful as it gives us a representation of how the aircraft will fly. Gliders have very low wing loading and can fly slowly and efficiently, whereas fast stunt planes have much higher wing loading, so they need to fly faster to generate sufficient lift. In order to make your drone as efficient as possible the lower the wing loading the better. However, you should make sure the wing loading is below $20oz/ft^2$ (ounces per square foot of wing area) for reasonable performance.

To find the wing loading, the first thing to do is to estimate the total weight of all the equipment your drone will be carrying. Then you calculate the total wing area of the aircraft and divide this by the weight to find the wing loading value in ounces per square foot. So the first step is to calculate the wing area, which is the span multiplied by the wing chord (width of the wing):

$$\text{Wing area (square inches)} = \text{wingspan (inches)} \times \text{wing chord (inches)}$$

As an example we'll use an aircraft weighing around 4.5lb, with a total wing area of $550in^2$. To find the wing area in square feet divide the square inch area by 144, to give 550/144 = $3.82ft^2$. You also need to convert the weight of the aircraft into ounces, so multiply the weight in pounds by 16 to give 4.5 x 16 = 72oz.

Now that we have the correct units we can find the wing loading by simply dividing the weight by the wing area:

$$72 \div 3.82 = 18.9 \text{oz per square foot}$$

A quicker way to calculate this would be to use the aircraft weight in pounds and the wing area in square inches and put the values into the following equation:

$$\text{Wing loading} = \text{weight (pounds)} \times 2{,}304 \div \text{wing area (square inches)}$$

Now that we know our aircraft has an acceptable wing loading value we can proceed to choosing a suitable motor. As a rule of thumb we need to have a motor power-to-weight ratio of around 80–120 Watts per pound (W/lb). Although your aircraft will fly with less powerful motors, having at least 100W/lb power will ensure it can produce enough thrust to climb fairly quickly, making the take-off process easier and requiring less space.

So, continuing our example, our aircraft weighs 4.5lb so we'll need a motor capable of producing 450W for a sporty/acrobatic FPV flying style. Alternatively we can use two motors that produce 225W each if our aircraft will support that.

Power-to-weight ratio (W/lb)	Aircraft type/flight style
50–80	Powered gliders and basic non-acrobatic aircraft.
80–120	More sporty and acrobatic-style flying.

Looking through a motor thrust table, check the maximum power for the various propeller and battery voltage combinations to find if any produce at least 450W of power. (This is slightly different to how we choose a multirotor motor/propeller value, based on the actual thrust value at 50% power.) Once you've found an initial motor/prop that's capable of producing this you can refine your entire selection by starting again with the new weights etc.

There are, however, more accurate methods of calculating what motor and propeller you need, but this can get rather complicated quite quickly, as you need to calculate the flight dynamics of your aircraft to find the flying speed based on aerodynamics and then use that to calculate the thrust required and choose your motor in a similar fashion to multirotor motors, based on required thrust. But doing this involves fairly detailed methods that are beyond the intended scope of this book, though if you want to take this approach there are resources and software tools available on the Internet showing how it's done.

Choosing an optimum motor operating point

Once you have an idea of the motor you need and a good idea of the other equipment you'll use on your drone, you can refine your motor selection by looking at the efficiency column on the thrust data table. This value is represented by g/W, or grams divided by Watts, which gives a representation of how much thrust the motor produces for the power it uses. The higher the g/W value the more efficient the motor is, and you'll notice that at lower thrust values this value is higher.

The table below considers an example of building a quadcopter with a total weight of about 3kg. We need to choose motors capable of creating at least 3kg of thrust at approximately 50% throttle, therefore each motor will need to produce at least 750g of thrust. Here we've chosen a generic motor capable of this and selected a few lines from the thrust test table at various battery voltages and propeller sizes:

As you can see, by using a higher voltage battery (a 14.8V four-cell instead of an 11.1V three-cell LiPo) the same motor can produce much more thrust if using the same 15*55-inch (or 15 x 55-inch) propeller, which could result in your quadcopter being much more responsive and faster. But by using a higher voltage battery you can still produce the required thrust with a smaller propeller, which might be useful if the frame you're using isn't big enough to accommodate 15in propellers. To get the most efficiency it would be best to use a three-cell LiPo and a 15in propeller in this case. However, it's worth noting that you can

BELOW Example of a few lines extracted from a generic brushless motor thrust table, showing the motor's performance specifications when using different-voltage batteries and different-size propellers. Please note that this isn't a complete thrust table for a motor, as a typical one would contain many more lines.

Voltage (V)	Propeller size	Current (A)	Thrust (G)	Power (W)	Efficiency (g/W)	Throttle (%)
11.1	15*55	6	690	66.6	10.4	43
		7	750	77.7	9.7	51
		8	810	88.8	9.1	58
14.8	15*55	6	800	88.8	9	29
		8	990	118.4	8.4	39
		10	1130	148	7.6	49
	13*4	6	750	88.8	8.4	52
		7	850	118.4	8	60

use other combinations to still build a drone capable of flying.

Matching an ESC to your motor
Once you've chosen the required motor and a desired operating point, you'll need to look at the thrust data table to see the maximum current draw. Once you know this value you'll need to ensure that the current rating of your ESC won't be lower than the maximum current draw of the motor. It's unlikely that the maximum current draw will exactly match your ESC, so it's best to round up so that your ESC will definitely be able to handle the required current draw from the motor. So if your maximum motor current draw is 14A you should use an 18A or 20A ESC. Using an ESC that can handle much more current is still allowable, but using a 50A ESC when your motor only draws a maximum of 14A is overkill and will just add extra weight and cost to your drone. It'll also use slightly more battery power.

Choosing your battery
Ultimately in order to increase flight time you'll need to use the highest capacity battery possible, though the weight of the battery will increase with its capacity. To get decent flight times you should try to keep the overall weight of your drone equipment, including the frame, as low as possible. That way the remaining weight can be used to accommodate a bigger battery. Looking back at the quadcopter example, we have motors that can lift a 3kg drone comfortably, and the combined weight of all the equipment and frame comes to 2kg. That means we should try to get the biggest battery possible that doesn't violate the maximum take-off weight of 3kg; so we can use a battery weighing around 1kg.

For fixed-wing drones you should use a battery that's as big as possible but still ensures that you don't exceed the wing loading value of around 20oz/ft^2. Lower wing loading values usually mean your aircraft will be easier to take off and fly.

Discharge rate
Other than making sure the battery isn't too heavy you need to ensure that it'll be able to provide enough current to your motors fast enough. Once again we need to refer to the thrust table to see the current draw of each motor then add them all together to get the overall current draw of your drone. For added safety you should use the maximum current draw of the motors at 100% throttle. For the chosen propeller and battery voltage of 11.1V the quadcopter motors have a maximum current draw of about 14A each, or a total of 56A. You don't really consider the power draw from other equipment such as autopilots, as the draw from these is minimal; unless you're using any specialist gear you usually need to add only 1A. If you want to be more accurate you can add together the power draw of all your equipment.

Current draw (A) = battery capacity (Ah) x battery discharge rate (C)

To calculate the maximum current draw of a battery you simply need to multiply the discharge rate (C rating) by the capacity in Ah. Let's say we decided to use an 11.1V, 10,000mAh LiPo with a discharge rate of 10C for our quadcopter. The maximum continuous current draw of the battery is 10Ah x 10 = 100A, which is well within the range of the 56A current draw of the motors.

Estimating the flight time of your drone
This is easier to do for multirotors than it is for fixed-wing aircraft, as you can exactly estimate the amount of thrust you need to hover based on the weight of your multirotor. With fixed-wing aircraft it's more difficult to know the exact thrust you'll use to fly without doing some more advanced calculations based on the aircraft's aerodynamics. But to estimate the flight time you simply need to divide the power the motors use to fly by the total power the battery provides. The details on how to find these values are discussed below.

To estimate how much power the motors consume to enable your drone to fly we need to consult the motor thrust data table. Because we know the weight of the drone and how much thrust the motors need to produce we can use that to find the g/W value for the required thrust to fly. As an example, consider a quadcopter that has a weight of 2,000g (2kg). Since the drone

ABOVE A quadcopter drone has a finite flight time that's largely dependent on the efficiency of its motors.

is a quadcopter with four motors, each motor will produce around 500g of thrust; so using the motor thrust data table, we find the g/W value when the motor is producing 500g of thrust. Let's assume the motors have an efficiency of 7.5g/W to produce the required thrust for the quadcopter to hover. So to calculate the power consumption of the motors we simply need to divide the thrust value by the g/W of the motors (2,000 ÷ 7.5 = 266W). Alternatively, some thrust data tables already provide the power consumption at various thrust values, and by looking at the table we'd see that the motors have a power consumption of 66.5W at 500g each; so the total power consumption of all four motors is 66.5W x 4 = 266W.

Now we need to find the total power in the battery. To find this value you simply multiply the battery capacity in Ah by the battery voltage. In our example we're using a 5,000mAh three-cell (11.1V) LiPo battery, so the total power is 5 x 11.1 = 55.5W.

In order to find the flight time in minutes we simply divide the two values and multiply by 60:

Total battery power (W) ÷ motor power draw (W) x 60 = flight time (minutes)

In our example the flight time will be:

55.5 ÷ 266 x 60 = 12.5 minutes

Although this value isn't perfectly accurate it'll give you a close estimate of how long you'll be able to fly with a fully charged battery, before you buy all your equipment.

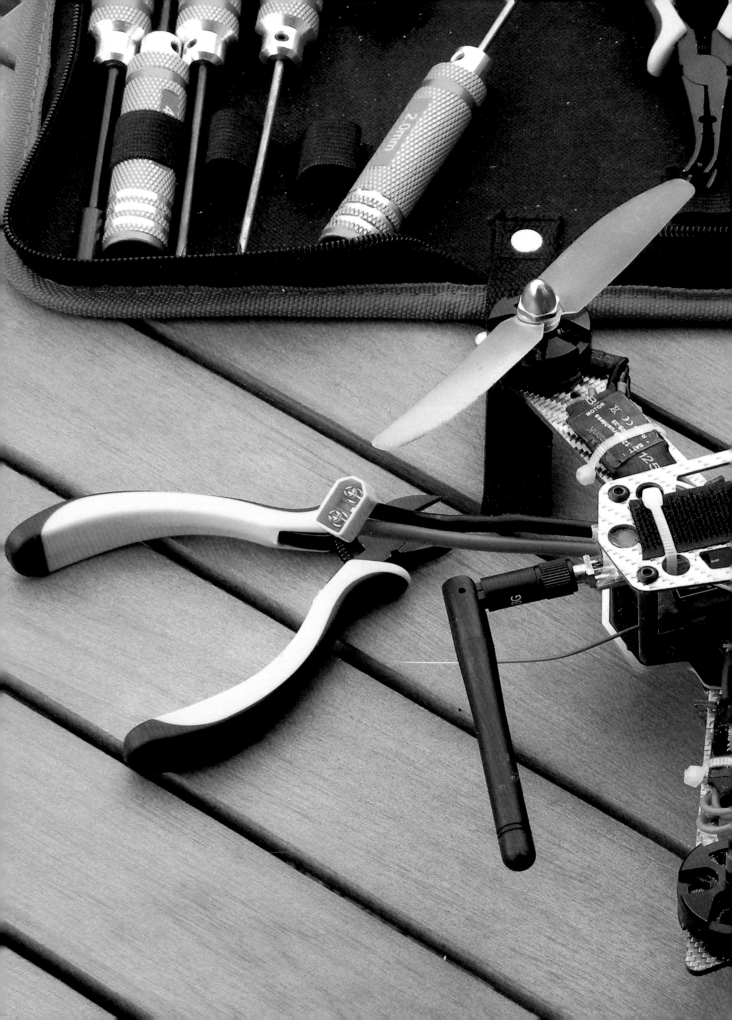

CHAPTER 5
DRONE BUILDS

(Sam Evans)

In this section we'll go through the process of selecting the various components for a drone and putting them together. When it comes to building a drone the exact specifics may differ somewhat, as each frame has its own unique assembly steps, but the accompanying instructions will show you how to construct it. So rather than go through the entire assembly process in this section we'll focus on some key building techniques and best practices that I've learned along the way, to ensure that your drone is built properly and safely. These guidelines are applicable to every drone build.

If you ever need any help or advice during a drone build, you can always ask on the community forums at www.dronetrest.com where the author, or another member, will be glad to help.

Aerial photography drone

This section will cover the process of building a versatile multirotor that can be used for aerial filming purposes. If you're new to drone building it would be best to start with a mini quadcopter drone, as it's cheaper and easier to build compared to a fully featured aerial photography drone.

Choosing the components

As previously discussed, you should first decide what you plan to use the drone for, as this will impact on the equipment you'll need to use. So since you want to be able to capture aerial photographs and videos you'll need to build a drone that can carry a camera and gimbal.

Autopilot/payload

Now that you've decided what the drone will be used for, you can look through various flight controller specifications to select one that has the features you require. For photography purposes I'd say the most important feature of a multirotor would be its ability to hover in position automatically. Having the ability to fly autonomous missions can be useful but isn't essential. However, some autopilot systems have extra features such as the ability to follow you automatically or circle around an object, and these can make it easier to record better videos. Another factor to consider is the comparative size and cost of different systems.

In this build I've chosen the Mini Unmanned Hawk autopilot kit, as it's small but has loads of autonomous features since it uses the open source ArduCopter software. The nice thing about this kit is that it also includes all the necessary components, such as GPS module and telemetry modules, for the autopilot system.

As we need a camera this is the next thing to choose. Maybe you already have a camera that you want to utilise, in which case the next steps would be to choose the required motors, frame and gimbal for that camera. However, to keep things simple we'll design the drone to carry a GoPro camera. Because the GoPro is light and small we'll be able to use a quadcopter design, since we'll probably only need four motors to produce enough lift.

Once you've selected the camera you want you need to find a gimbal for it so that the footage will be smooth. In terms of GoPro-size gimbals there's literally hundreds to choose from, but some work better than others, so it's worth doing a quick search on YouTube for the gimbal you're considering to see some sample footage. For other larger cameras there might be less choice, but the important factors to look out for are that the gimbal will support your camera weight and has enough adjustment points that you can get the camera properly balanced. If a gimbal is designed for a specific camera this might not be so important, as the mounting position is already optimal, but if it's a general gimbal then having sufficient adjustment points is critical to getting the best performance from the camera.

Estimating the weights of your gear

In order to choose a motor that's powerful enough to lift everything you need to know the approximate weights of all of your equipment. However, this can be difficult, as you haven't yet chosen the motors, battery or frame, so at this point you can only estimate these weights. As you get more familiar with various pieces of gear this will become easier, but if you have no idea the best place to start is by scouring the Internet for other pilots who've shared details of their own drones. If you're lucky you might even find someone who built a drone using exactly

the same camera gear as you're going to, which you can use as inspiration.
Things we know the weight of:

- Autopilot system – about 140g
- GoPro camera – 75g
- GoPro gimbal – 190g

Things we must estimate the weight of:

- 4 x motors/propellers/ESCs – 500g (on average most motors/ESCs come to about 125g each)
- 10,000mAh battery – 800g (we want to be able to fly a long time, hence the large battery)
- Quadcopter frame – 800g

So the approximate total weight of your drone will be about 2.5kg. Therefore to hover, all the motors combined will need to produce a total thrust matching this figure.

Motors and propellers

Now that you have an idea about what equipment you need to carry on your drone you can start looking for motors that are powerful enough. The rule is that you need to have a power to weight ratio of 2:1, so the motors need to be capable of lifting twice the total weight of the drone. Since we're building a quadcopter each motor will need to produce 1.25kg of thrust:

2.5kg x 2 ÷ 4 motors = 1.25kg per motor

As discussed in previous chapters, now that you have an idea of the thrust the motors must produce you can consult the thrust tables to find a suitable motor and propeller combination. Once you've found a few suitable motors you can compare their efficiency at the actual thrust value for hover (625g), with the higher values being better as they'll allow you to fly for longer.

I've decided to use the MT 3510 motors from Unmanned Tech Shop. These are capable of producing about 1.8kg of thrust each, using 15in propellers and a 4S battery. You might think that this motor is way too powerful for our quadcopter, which is probably true, but if you look at the efficiency of the motor when it produces around 600g of thrust you'll see that it's very good, at about 10g per Watt. In addition to that, having powerful motors means that you can always upgrade the quadcopter to carry a larger camera should you want to in the future. Comparing it with a smaller motor, the MT 2216 has a g/W of just 6.1 at around 625g compared to 10g/W for our chosen motor. Having a better efficiency means that you can fly your drone for longer, enabling you to capture more footage.

Here are the specs for our chosen motor at the approximate hover point:

Voltage (V)	Propeller size	Current (A)	Thrust (G)	Power (W)	Efficiency (g/W)	Throttle
14.8	15*55	4	590	59.2	10	Hover
14.8	15*55	20.2	1,780	299	6	Maximum

Choosing an ESC

Now that we've selected a motor and propeller capable of lifting our drone we can look at the thrust data table to find the maximum current pull from the motor, to ensure our chosen ESC is big enough. The max current draw of the MT 3510 motor with the 15in propeller is 20.2A. It's always better to get an ESC slightly bigger than the maximum current draw to avoid it becoming too hot, so I decided to use a 30A ESC with this drone. I also made sure to choose an ESC that has multirotor-specific firmware.

Multirotor frame

At this point you can start to look at frames that have features you like the look of. Comparing maximum take-off weights will quickly show you if a frame is suitable to carry your camera gear. Usually the more weight the frame supports the heavier it is, as it needs to use stronger materials and additional reinforcement.

I decided to choose the U580 Pro quadcopter frame, as it includes several built-in features that make it ideal for use as an aerial photography drone, while having arms that fold down like an umbrella means it'll be easier to transport. Its main features for photography include a nice vibration-isolated loading boom system on which you can mount your extra equipment, such as gimbal and battery. Retractable landing gear is also included, enabling it to fold upwards out of view of the camera when you're flying. The weight of this frame is also just 640g, which is lighter than our estimate.

Choosing the battery

Now that we know the actual weights of everything and how much the motors can lift, we can make a better estimate of the size of battery we need. The four motors together are capable of lifting a total of 7kg, so in order to have sufficient power the total weight of the quadcopter must be around 3.5kg. After subtracting the weights of the motors, camera and frame, etc, the weight we have left is what's available for the battery.

In terms of the discharge rate, for aerial photography purposes we'll be flying smoothly rather than aggressively, so it's best to have a discharge rate as low as possible as this allows for a lighter battery. Looking at the maximum current draw of the motors/prop combination we can see this is 20.2A. Multiplying that value by four to get the total current draw from all the motors gives a value of 80.8A. To find the maximum current draw from a battery, as discussed earlier, we multiply the capacity in Ah by the discharge (C) rating. Our chosen battery has a capacity of 10,000mAh (10Ah) and a discharge rate of 10C, so the maximum current draw is 100A. Our maximum motor current draw of 80.8A is well within this figure.

Weighing it all up

Now that we've made initial selections for all our equipment we need to ensure that it'll all fit on to the frame – if not we'll need to revisit some choices. We can also add all the weights together to ensure the quadcopter will be able to fly at around 50% throttle:

- Payload weight (camera, gimbal, autopilot) – 405g
- U580 Pro frame – 640g
- 4 x MT3510 motors, ESCs and propellers – 550g
- Four-cell 10,000mAh 10C LiPo battery – 805g

Total drone weight will therefore be around 2,400g. However, we should add an extra 100–200g to account for additional miscellaneous equipment such as battery straps, zip ties and other cables.

Estimating the flight time

Using the equation discussed in the previous chapter, we can get an estimate of the flight time by dividing the motor power usage by the battery power.

Looking at the thrust table, the motor efficiency at hover is about 10g/W. To calculate the power used we divide the weight of our quadcopter (2,600g) by its g/W figure (10) efficiency, which is 260W. The total battery power is found by multiplying the capacity (10Ah) by the voltage (14.8V), giving a value of 148W.

By dividing the motor power by the battery power and multiplying by 60 we can get an estimate of the flight time in minutes:

$$148 \text{ (battery power)} \div 260 \text{ (motor power)} \times 60 = 35 \text{ (minutes)}$$

So we'll be able to fly our drone for about 30 minutes on a fully charged battery, which is great. In reality, however, your flight times will be lower, as batteries wear out with time and other factors such as temperature and wind could mean the motors have to work harder. Also, the thrust data tables are generated under static conditions, so the calculated efficiency values are often better than they'll be in reality. Nevertheless, this calculation does provide you with an estimate of what sort of times you can expect.

Building the quadcopter

Now that we've done some calculations to ensure that the drone will be able to fly properly, we can start to build it. I won't go into the specifics of building this exact frame, as its owners' manual will show you how to do this, and each frame is unique. Instead I'll focus on some tips and things to consider when attaching your gear, so as to ensure you build a high-quality drone.

The first thing I do when I start to build is to plan out where everything will go on the frame. This way I know in advance which wires I need to shorten or lengthen, what sections I need to pass wires through and so on. However, I admit that after planning to put gear in a certain place I often find halfway through the build that there's a better location, which means I have to re-solder a bunch of things – which can be frustrating, but in the long run makes for an optimised drone set-up.

1 A word of advice before you start – use a thread locker on all the screws. Vibrations are inescapable on a quadcopter and over time can cause screws to loosen, which could result in something catastrophic occurring during a flight. To reduce the chances of this you should use thread locker or Loctite on *all* your screws.
(All photographs in this Chapter by Author)

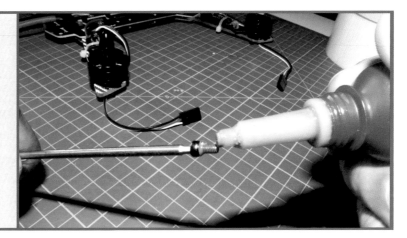

Motor and ESC connection

2 I often start a build by mounting the motors on the quadcopter's arms, as there's only really the one place for them to go. But make sure to double-check the motors' orientation, because certain autopilot systems require motors to spin in specific directions; it's therefore worth checking this first to ensure that each motor is mounted on the correct arm.

Some frames have space in the central section for the ESCs, some allow space to mount them on the arms, and others have a space on the motor mounts. The U580 quadcopter we're building has space to mount them directly under the motors, which is a great place for them as the airflow from the spinning propellers helps to keep them cool.

3 Motors and ESCs both usually come with bullet connectors attached, so they're easy to connect. As a tip, whenever you're using bullet connectors make sure that they're connected firmly, and ensure the soldering is done properly by testing the motors on the ground before flying. On occasion I've had a motor fail due to faulty soldering on a bullet connector, so it's something worth taking a few moments to check before you go out and fly. An easy way is to wiggle the connectors gently with the motor spinning; if there's a loose connection it will probably stop.

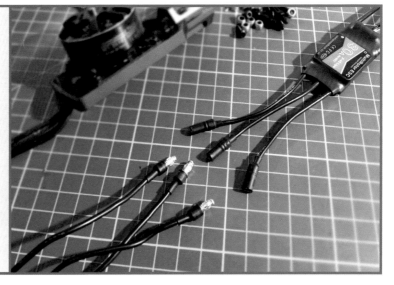

DRONE BUILDS

4 Motor and ESC cables are usually too long and need to be shortened to a couple of centimetres, which is sufficient for the ESC and motor wires to be soldered together. Using wire cutters/strippers makes it relatively easy to remove the insulation from the ends without damaging the wires themselves. At this point you could solder some new bullet connectors to the ESC and motor, but in this build we're going to solder the motor wires directly to the ESC as this is the most permanent and secure method.

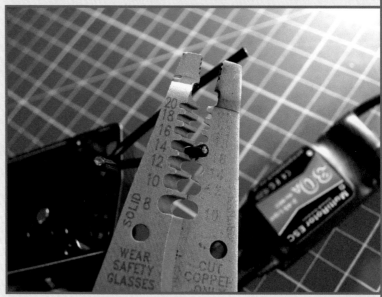

5 After all the wire ends have been stripped apply a small blob of solder to each of them. This makes soldering the two wires together much easier. If you aren't familiar with soldering it's a good idea to watch a few tutorial videos on YouTube, as there are plenty that will show you how to become proficient. My only advice would be to invest in a decent soldering iron, as some of the cheaper ones aren't powerful enough to maintain their heat, which makes connecting thicker wires almost impossible.

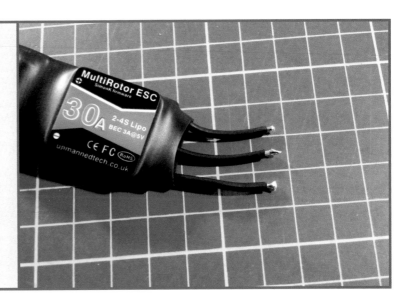

6 Using clamps makes the process of soldering wires much easier, as it frees up both hands. Something that you need to remember – and that I constantly forget – is to put heat shrink tubing on the wires before you solder them together. When heat's applied to this it shrinks tightly over wires, forming a neat connection. If you forget you'll have to undo the soldering, slide the heat shrink over the wires and then solder them again.

7 You'll recall from a previous section that a motor's spin direction is dependent on how you connect its wires, so before you permanently solder the wires together you must ensure that you're connecting them correctly so that the motor will spin in the desired direction. With two motors you'll need to cross the outer two wires so that they spin anticlockwise, while on the other two the wires should not be crossed (as in the picture), so that they spin clockwise.

8 Once you've soldered all the wires together slip heat shrink over the exposed metal to insulate them.

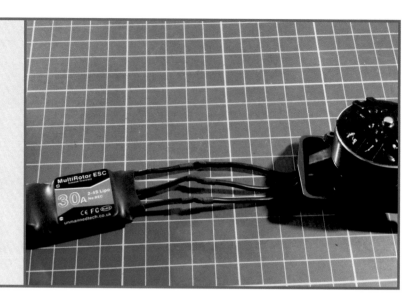

9 Repeat this entire process for the other three motors, remembering to cross over the outer two wires on two of the motors to ensure they spin in the opposite direction. Your flight controller documentation will define the required motor order and spin direction.

10 Next you need to check that the ESC power cables are long enough to connect to the central part of the quadcopter frame. In order to assess the required length lay the cables out next to the carbon fibre arm of the quadcopter. If they're not long enough you'll need to extend them.

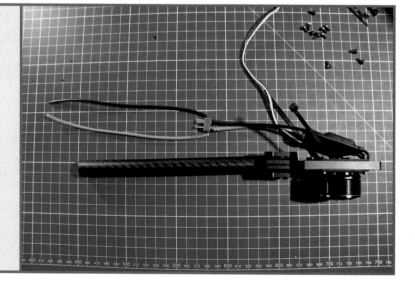

11 Before you solder on your extensions make sure that the wire you're using can handle the required current draw. The easiest way to do this is to check the gauge of the wire on the ESC. In this case the ESC has a 16AWG wire (the spec is helpfully written on the cable itself), so you need to use the same gauge to ensure it won't overheat.

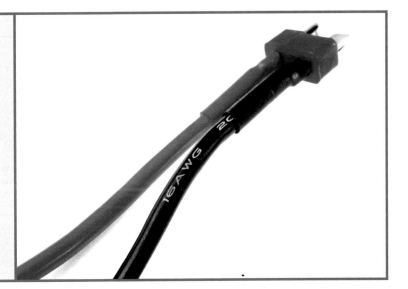

12 Repeat the process for all four ESCs, soldering extra wire to the end of each cable and insulating the connection with heat shrink.

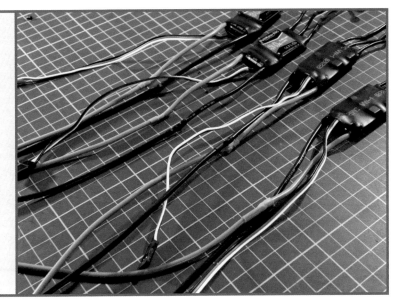

13 Next mount the ESCs to the underside of the motor mounts using zip ties. When securing the ESCs you can fold any excess cabling between the motor and ESC so that it fits neatly underneath.

Routing wires through the arm tubes

14 A neat way to hide the ESC and motor cables is to route them through the arms of the quadcopter if they're big enough. However, if you do this ensure that the wires don't get pinched by any of the mounting screws passing through the centre of the arm. If a screw catches on a cable it could cause a short circuit or break the cable completely, so take your time when attaching the arms to ensure this doesn't happen.

Connecting the ESCs to the power distribution board

15 In order for the ESCs to get power from the battery they're all soldered to the power distribution board. The U580 frame is nice as its bottom plate has a PDB built into it, which makes things much easier and neater in terms of power distribution. Simply solder the ESC's red wires to the + (positive) solder pads on the PDB, and the black wires to the - (negative) solder pads. When soldering wires to the PDB it helps to add a small blob of solder to the pad itself first, and then solder the tinned wire end to that.

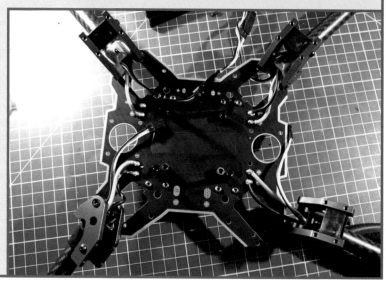

16 Because it's never a good idea to have exposed live wires anywhere on a quadcopter, you should stick some electrical tape over the solder pads to insulate the connections. This will prevent anything on the ESC from accidently causing a short circuit.

17 All the ESCs are now connected to the PDB, which essentially means that all the red wires are connected together and all the black wires are connected between the ESCs. By connecting power from the battery to the remaining positive and negative pad on the PDB it will power all the ESCs.

Since the Mini Unmanned Hawk autopilot on this drone has a power module that measures the voltage and current from the battery this needs to be connected between the battery and the power distribution board. When it's soldered direct to the PDB the battery is connected to the power module. The module has an input and an output for the electricity to flow in from the battery and out again. When connecting the power module you need to ensure you connect the output side to the PDB.

The two remaining exposed solder pads next to the arrow will be used later to power the gimbal and FPV gear.

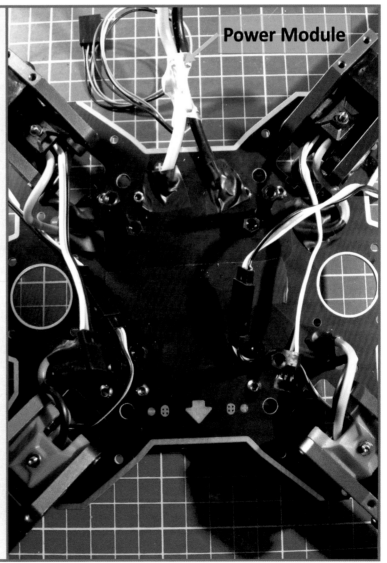

Attaching the autopilot

Now that we've completed the major wiring of the quadcopter's power system we can move on to connecting the autopilot.

18 Since electricity flows through the bottom plate of the frame you mustn't mount equipment such as compass sensors to the PDB board – electricity generates its own magnetic field that would interfere with the sensor readings. The Mini Unmanned Hawk has a compass sensor built in, but it deals with interference by using a secondary sensor that's built into the GPS module and mounted on a mast some distance away. So depending on what autopilot system you're using this is something of which you need to be aware.

To ensure it provides the best possible readings it can the autopilot needs to be placed on an anti-vibration mount consisting of two fibreglass plates separated by soft rubber balls that absorb any vibrations. Installing these balls can be quite tricky, but with a pair of blunt tweezers you should be able to gently squeeze them through the mounting holes. Be careful not to pierce or tear the balls.

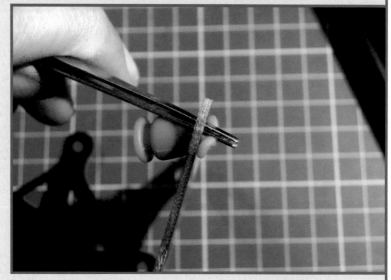

19 Screw the vibration mount to the quadcopter's central frame and attach the autopilot to it with double-sided foam.

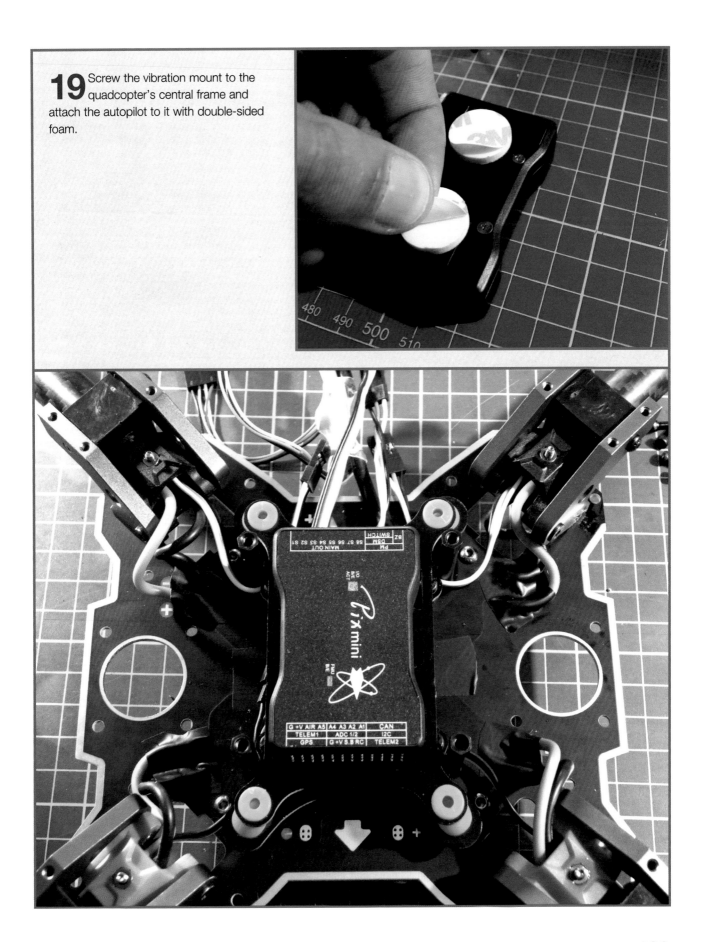

20 The next step involves connecting the ESCs to the autopilot. For this you'll need to refer to the diagram that comes with your autopilot instructions, as all autopilots require the motors and ESCs to be connected to specific outputs on their boards. In the case of this autopilot S1 goes to the front-right motor, S2 goes to the back-left motor, and so on. It's critical that you connect the ESCs to the correct outputs or your drone won't fly. You should also take care to ensure the three-pin ESC connectors are connected the right way round, with the signal, 5V and ground going to the correct pins on the autopilot. In the case of the Mini Unmanned Hawk the signal pin is on the top. In this build I chose to use one ESC with a battery eliminating circuit, while the others don't have a BEC so it isn't necessary to remove the red wires from them (as discussed in the ESC section in Chapter 4).

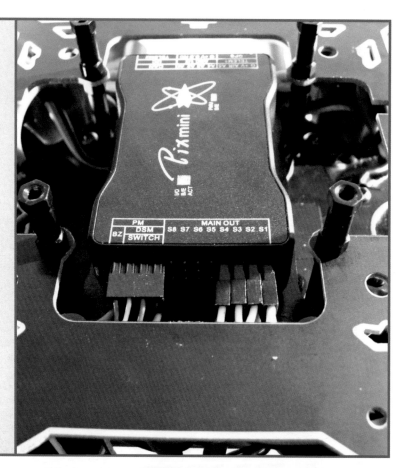

21 This particular autopilot doesn't include a neat way to mount its safety switch, so you'll need to drill a hole in the top plate to do this. The safety switch is a clever feature that disables the motor's output until you press and hold the switch down for a second. This prevents the motors turning on accidentally if you bump the throttle stick of your R/C transmitter.

Before drilling any holes in the frame it's essential to ensure the plate doesn't have a built-in power distribution board, as drilling through that could damage the circuit built inside. The top plate of the U580 frame doesn't have any built-in circuitry so this isn't a problem.

Installing landing gear and ancillaries

22 Now is a good time to mount the quadcopter's retractable landing gear. This includes a small control box that extends and retracts the gear. These particular landing gear can be powered with a 2S to 6S LiPo. Since we're using a 4S LiPo the red and black power cables should be connected to the power distribution board to draw power directly from the battery (this is where we use the two spare solder pads that were mentioned earlier). There's also a black and white connecter used to tell the landing gear to extend or retract, which needs to be connected to one of the autopilot outputs. The exact one will be mentioned in the autopilot instructions.

Alternatively if your autopilot doesn't support this function you can connect the landing gear control connector directly to a spare channel on your receiver, enabling you to bypass the autopilot and control it manually via a switch on your R/C transmitter.

23 As you continue to connect the rest of the autopilot's equipment, such as the telemetry kit and the GPS/compass module, your quadcopter starts to look more and more like a drone. Always be sure not to plug in a connector the wrong way round. Some autopilot systems use connectors that only fit in one way round in order to prevent this, but others, including the Mini Unmanned Hawk, don't have this feature due to their compact size. So it's up to you to double-check that everything is connected correctly. Specific details of where and how to connect everything are often included in the form of diagrams in autopilot system documentation.

Attaching the R/C receiver and GPS

24 The next vital piece of equipment to attach is the R/C receiver. To help it get the best possible signal a 3D-printed antenna mount is used to position the antennae at exactly 90° to each other. Like most of the other electronics the antennae are mounted using double-sided foam tape. The receiver we're using has the ability to output a PPM signal so we only need to connect a single servo wire between the autopilot and the receiver and don't need to use a PPM encoder.

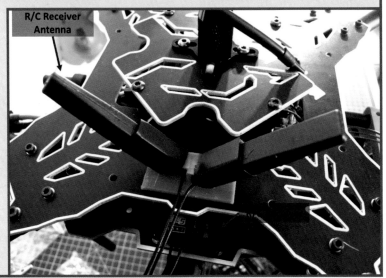

25 Mounting the GPS/compass module on a mast helps minimise interference from other electronic sources and ensures a strong signal, but you must make sure that the built-in compass is mounted facing forwards. Most modules have an arrow on them so that you can easily identify the front. The autopilot instructions will provide further details.

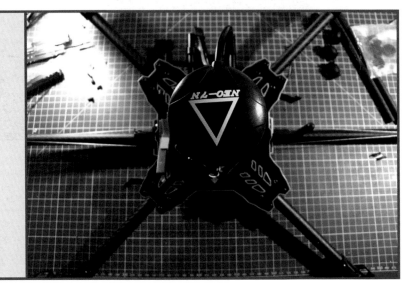

Attaching the battery

Next we need to mount the battery to the quadcopter's load booms.

26 Because I plan to use a gimbal with our quadcopter having the battery on the bottom could potentially cause issues, as there might not be enough space underneath for both gimbal and battery. For this reason some multirotor frames are designed to have the batteries mounted on top of the central plates.

When choosing a battery it's a good idea to get one with the same type of connector as the power module so that you can just plug it in, since with most hobby drones you turn them on by plugging them in and off by disconnecting the battery. To ensure the battery doesn't slide around you should add some Velcro to both the battery and the battery mounting area. This isn't enough to hold the battery but it will prevent it from sliding around. The battery itself is mounted with a Velcro battery strap.

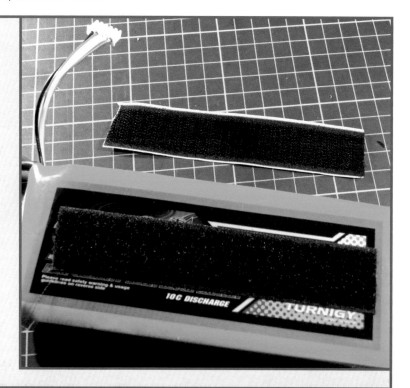

27 Apart from having enough space to mount both battery and gimbal you also need to ensure that the quadcopter's centre of gravity is approximately central. You can easily check this by balancing it on your fingers placed just beneath the middle of its central section. If it leans to one side you'll need to adjust the positions of heavy equipment such as the battery or gimbal until it stays approximately level.

If this is the first time you've built a drone it would be a good idea to stop at this point, set up the autopilot system as per its accompanying instructions and start flying. For the reasons discussed earlier you should always start simple, since adding more equipment such as a gimbal or FPV equipment just adds extra complexity. You should only add extra stuff once you're confident about flying your drone.

Mounting the gimbal

28 Like the battery, the gimbal is attached to the load booms. However, we need to provide power to the gimbal for it to work. Since the GoPro gimbal we've chosen can be powered with a 2S to 6S LiPo we can once again solder its cables directly to the U580 power distribution board.

Most gimbals allow you to adjust the pitch or yaw angles during flight so these cables need to be plugged into your autopilot output connectors. The locations of these connectors are specific to individual autopilots, so follow the instructions included with the autopilot on how to connect and enable the gimbal outputs.

Adding FPV equipment

29 In order to get a live video feed from the GoPro camera we'll be mounting on our drone we need to add an FPV video transmitter. The GoPro camera uses a special USB video cable, which plugs into its side. The other end is connected to the FPV transmitter.

Since the transmitter can be powered by 2S all the way up to 6S LiPo batteries we can once again solder its power cable directly to the PDB. The FPV transmitter that we're using has a built-in power filter, so powering it from the same circuit as the motors/ESCs should be fine. However, if you have problems with the video signal getting interrupted when the drone's motors turn on it would be best to add a dedicated power filter between the PDB and FPV transmitter/camera.

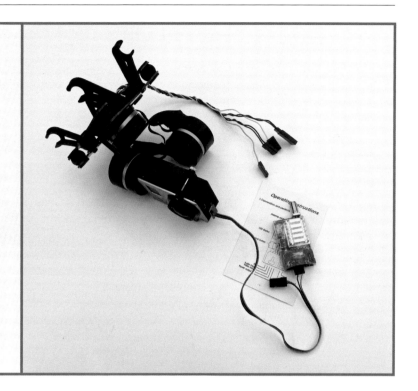

Final touches

30 Now that all the equipment has been mounted on your quadcopter, the software on the autopilot needs to be set up. Once again, this is very specific to the autopilot system you're using, but every system includes instructions on how to set it up. Usually this will involve some initial calibration of the sensors, ESCs and R/C radio. You'll also need to specify various other parameters, such as the size and capacity of the battery you're using. Depending on the control system it might also be necessary to perform some PID tuning to ensure the best possible performance, as discussed in Chapter 4. Most flight controllers, however, provide adequate performance straight out of the box, so you don't always need to do this. If you ever have any issues or need help you can contact me, or anyone else in our community, for help at www.dronetrest.com.

Whenever you change any settings on your autopilot *always* remove the propellers to avoid any chance of damage or injury.

Mini FPV quadcopter

In this section we'll build a mini FPV quadcopter that can be used for racing. Racing quadcopters are designed to be as fast and as agile as possible and only fly for a few minutes, so efficiency and long flight times aren't usually the main concern. Mini FPV quadcopters are great fun due to their small size, which enables them to dart through narrow gaps between obstacles at high speed.

Choosing the components

Mini quadcopters are categorised by their size, such as 250 or 300 size. These numbers represent the distance in millimetres between the shafts/centres of the two furthest separated motors (*eg* front right to rear left).

Since we want to build a racing quadcopter, speed and power need to be our top priorities when choosing components. Unlike other drones, when it comes to mini FPV quadcopters it's usual to start by choosing the frame. This is because some racing categories specify a maximum frame size. However, depending on where you race or who you're racing with I generally prefer to enter races based on maximum prop size and battery voltage, as this is more fair. So in this example we'll take a slightly different approach by building our basic quadcopter on a 250-size frame, with a 3S LiPo and 5in propellers.

Choosing an autopilot

FPV racing quadcopters only need basic flight controllers so these are usually very small and lightweight. There are many to choose from, but I've selected a Flip32+ for our drone as it can run Cleanflight open source quadcopter firmware (which can also run on several other flight controller boards). The Flip32+ board includes a couple of extra sensors such as a compass and barometer, which make is easier to fly for beginners, as you can use features such as automatic altitude hold.

Estimating the weight

To find suitable motors we'll first need to establish the total weight of the components. We can then select the best, in terms of power, for 5in propellers.

- Flight controller – 15g
- R/C receiver – 15g
- 250-size quadcopter frame – 150g
- 4 x motors and ESCs – 200g
- 3S LiPo battery – 150g
- FPV camera and transmitter – 50g

The weight of our drone will therefore be around 580g, so we'll need motors that can produce at least twice the necessary amount of thrust, or about 1.2kg of thrust in total. However, if they can produce more we'll probably be able to fly faster.

Choosing the motors and ESCs

Since we require a total thrust of about 1.2kg each motor must produce about 300g. We therefore need to look through motor specifications and thrust tables to find one that matches this figure. In terms of miniquad motors in general most of the ones you'll see will be around the 1806 to 2208 size. I've decided to go with a 2,204–2,300KV motor, as it can produce a maximum of 310g thrust with a 5in propeller, which is sufficient. Furthermore, if we decide to use a 6in propeller at a later stage this thrust value goes up to 440g. This particular 2,204 motor can be powered by a 2S–3S LiPo and weighs only 25g. Looking at the specifications a 12A ESC will be sufficient.

Choosing a quadcopter frame

There are countless miniquad frames to choose from but they're all very similar, as they fall into the same 250-size class. Some are made from fibreglass, others from pure carbon fibre, which is stronger but more expensive. At the end of the day choosing a miniquad frame primarily comes down to what you like the look of!

I've opted for the Silver Blade quadcopter frame for three main reasons. The first is that it includes a built-in power distribution board to make soldering the power connectors easier. The second is that the frame itself is simple, which means it won't take very long to build, or to repair if I have a crash. Finally I like the look of the silver carbon fibre (although the frame is actually made from a fibreglass and carbon fibre composite). At 120g its weight is also slightly below our initial estimate of 150g.

Choosing a battery

Now that we've chosen the relevant equipment we can get a better estimate of the drone's weight, to see how much is left for the battery. Adding up the new weights shows that the quadcopter's total weight excluding battery is now 360g. Based on the motors we've chosen the 50% thrust level is 620g, so we need to find the best battery we can that weighs under 300g (ie 620g thrust minus 360g total weight).

An important factor when choosing a miniquad battery is to ensure that it can provide enough discharge current. This is more important than on larger quadcopters due to the smaller capacity. Therefore with typical miniquad batteries you usually need to find one with a discharge rate of at least 35C, depending on the capacity. I've decided to go with a 3S 1,800mAh LiPo with a discharge rate of 45C and a weight of 160g. The maximum current draw of this battery is 81A (1.8Ah x 45C), so since we're using four 12A ESCs that would draw a maximum of 48A we should be OK.

Since we'll have some spare weight available when we hover at 50% thrust we can either keep our quadcopter lighter in order to fly faster and for longer, or use that weight to add an on-board HD camera.

Building the mini FPV quadcopter

Before you start building your drone you should take the time to plan out where you'll mount all of your equipment, as it can be very frustrating to find when you get the last parts that you have to dismantle a bunch of things in order to get everything to fit.

Building the central frame

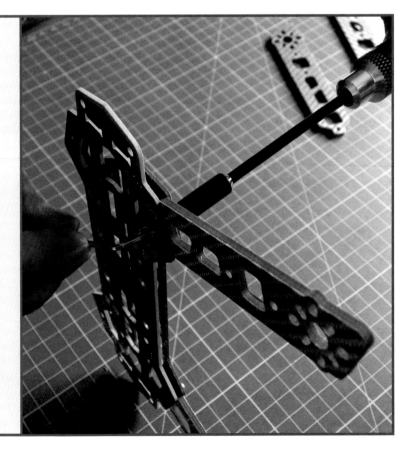

1 As with most parts, you'll usually – but not always – receive assembly instructions showing you how to build the frame, and it's best to follow these as they generally suggest the most practical order in which things should be put together. However, in the case of the Silver Blade frame no instructions were included, so I started by attaching the arms to the central power distribution board using nuts and bolts. As with all builds, make sure you use Loctite on bolts and screws to prevent things coming loose due to vibrations.

2 Once the main part of the frame is assembled we can move on to soldering the motors and ESCs together. The plan is to mount the ESCs on the arms of the quadcopter in order to keep the main platform free for other equipment. However, some mini quadcopter frames have the option of mounting the ESCs on the central plate if you wish. Before soldering the motors to the ESCs you should position them on the frame to see how long all the wires need to be, and cut or extend them as necessary.

Soldering the motors and ESCs

3 Once the wires have been cut to size you can proceed to solder the motors to the ESCs as on the aerial photography drone described earlier. Also, remember that two of the motors will need to spin clockwise and the other two anticlockwise, so you'll need to cross the outer two ESC wires on two of them.

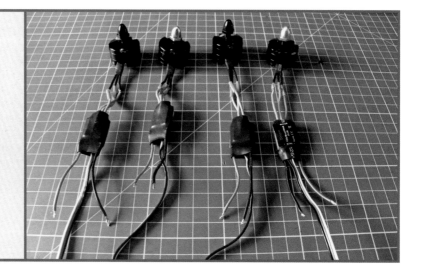

Mounting the motors to the frame

4 Once the motors and ESCs have been soldered you can attach the motors to the frame, remembering to check the spin directions and to use Loctite on the screws. Your flight controller instructions will tell you what order the motors need to be connected in and which direction they need to spin. In the case of the Flip32+ the front-left motor must spin clockwise and the front-right anticlockwise.

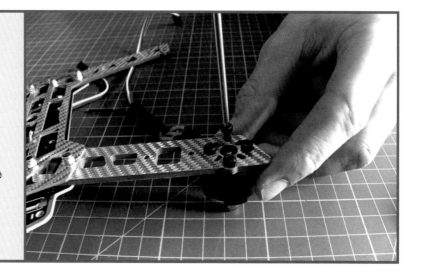

5 Once the motors have been attached to the arms the next step is to solder the ESCs to the power distribution board. The black wires (negative) of each ESC are all soldered to the negative solder pads on the power distribution board, and the red wires (positive) are soldered to the positive pads. Once the ESCs are soldered you should solder the battery connector cable to the appropriate pads on the rear of the PDB.

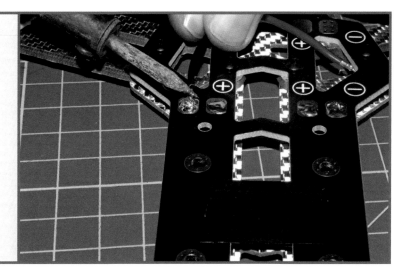

6 Once everything has been soldered you can fix the ESCs to the arms using zip ties. If the wires are slightly too long you can keep things neat by folding them under the ESC before you fix it.

7 Once the ESCs and battery connectors have been soldered in place you're ready to move on to attaching the other flight electronics.

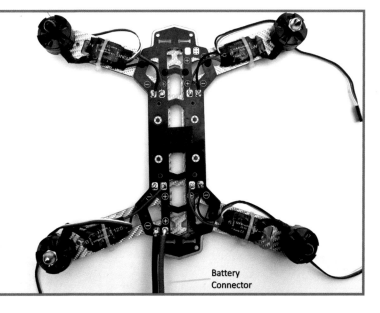

Battery Connector

Mounting the flight electronics

8 Before sticking down the Flip32+ flight controller make sure that it's correctly orientated, since most flight controllers need to face forwards. Some flight controllers have the ability to set the orientation within the software, so sometimes its easier to mount the board facing the side, so you can easily access the connectors, and then set the orientation correctly via the software. I used double-sided foam tape to attach it, as this has the added benefit of damping some of the quadcopter's vibrations. Once it's fixed to the centre of the board you'll need to connect the ESCs to the flight controller output ports. Refer to the instructions to see which ESC/motor needs to be connected to which pin number on the flight controller – make sure you connect them to the right output pins. If you're using ESCs with switching BEC it's best to remove the red wire from all but one ESC servo connector to avoid any power conflicts or interference.

At this point you should cover all the other solder pads on the PDB with electrical tape, to avoid the chance of a short circuit. It's also best to tidy the ESC servo cables by folding them neatly and taping them together.

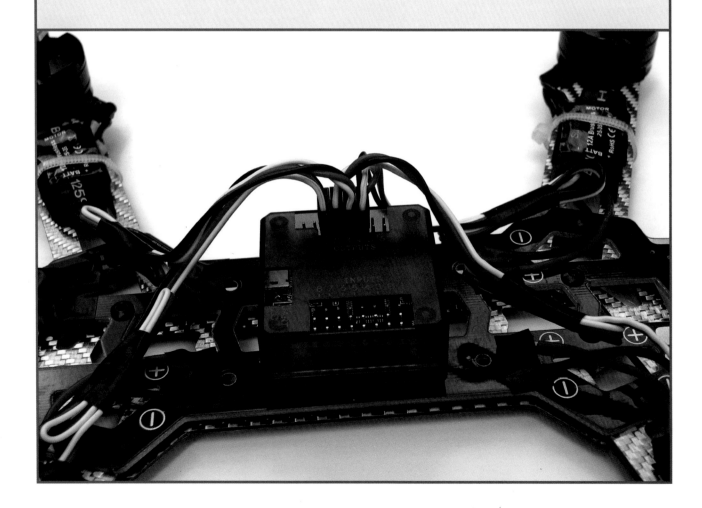

9 We can now connect the R/C receiver to the inputs of the flight controller. Again, the flight controller instructions should tell you which receiver channel goes into which input on the flight controller. In this example the R/C receiver can support PPM (which means it can be connected using just one servo cable), but for illustration purposes I'm using the PWM outputs so each channel needs to be connected with a cable. In order for the flight controller to provide power to the R/C receiver we need to connect one 5V and one ground cable to the appropriate pins on the receiver. Since the flight controller provides power to the receiver we need to use a single cable for each of the channels, and then one wire to provide power to the receiver and one wire to connect to the ground.

Now that the main connectors on the flight controller have been connected we can secure the R/C receiver to the back of

the frame. Since an R/C receiver is something you might also want to use on other vehicles it's best to secure it with Velcro, since this makes it quick and easy to remove. You might also need to remove it to bind it to the R/C transmitter when setting things up for the first time. It's best to attach the R/C antenna below the quadcopter to ensure it gets a reliable signal.

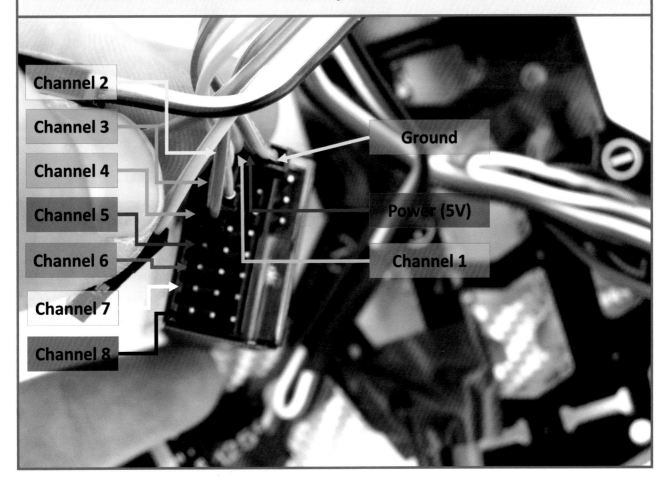

Attaching the FPV equipment

10 The plan is to power the FPV equipment directly from the balance lead connector on the battery, so we need to make sure the FPV transmitter and camera are able to handle the voltage of the three-cell battery before we connect them. If you experience lots of noise on the video signal with the motors running you might need to use a filtered voltage regulator between the battery and the FPV equipment.

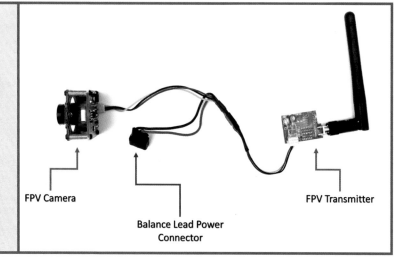

FPV Camera — Balance Lead Power Connector — FPV Transmitter

11 The first thing to do is connect the camera to the transmitter and solder a balance lead connector to the power input cables. The FPV camera is then mounted on the front of the frame between the top and bottom plates. Most people angle the camera upwards slightly so that when they're flying (and the quadcopter is tilted forward) it's closer to the horizon, otherwise they'd be perpetually looking at the ground and wouldn't be able to see any obstacles ahead.

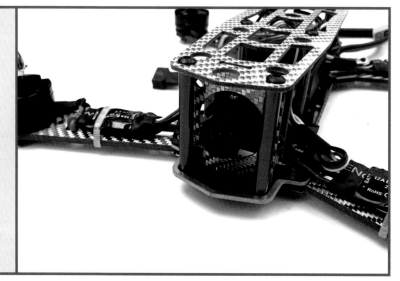

12 Secure the FPV transmitter to the underside of the top plate at the back of the quadcopter using double-sided foam and a zip tie. It's a good idea to set up the channel switches in order to set the appropriate frequency before attaching it, as this can be slightly awkward once it's been secured in place.

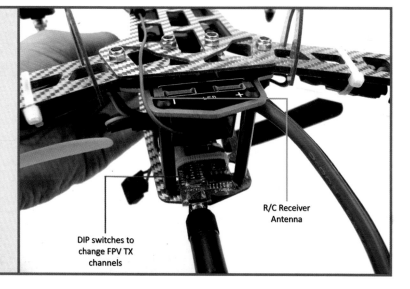

DIP switches to change FPV TX channels — R/C Receiver Antenna

Finishing up

13 Our quadcopter is just about finished. We just need to attach the battery to the top plate. However, you might also want to consider adding an anti-vibration camera mount if you plan to record HD video with a GoPro or a similar action camera.

The battery is simply attached to the top plate using some Velcro battery straps. To prevent it sliding about it's also a good idea to attach some Velcro to the battery itself. If you're not going to be using an action camera it's best to mount the battery in the middle of the quadcopter to ensure the centre of gravity is in the middle of the frame; but if you're going to be using an additional camera the battery will need to be mounted further back, as the camera will then offset its weight and keep the centre of gravity in the quadcopter's approximate middle.

Once everything has been installed you should give the quadcopter a once-over to ensure that all the cables are properly secured and have no chance of hitting any of the propellers as you're flying.

Now that the frame has been built you can proceed to configure the flight controller and set the flight modes via the flight controller software on your PC. This is a fairly simple and well-documented process; some flight controllers such as the CC3D even include a set-up wizard that guides you through the entire process – all you need to do is connect the flight controller to your PC via a USB cable.

Whenever you change any settings always remove the propellers in case something accidentally switches on, as you don't want to injure your fingers.

14 Among the other things people like to add to their mini quadcopters are LEDs, which help make them much more visible to other FPV pilots, particularly when flying in darker environments. They also look cool! The Silver Blade frame includes some LEDs that you can solder to the front and back of the power distribution board if you want to. Some flight-controller software, such as 'Cleanflight' (which we are using for this quadcopter) has the option to control RGB LEDs to make them act like brake lights, or turn indicators, so that other pilots can easily see what your your drone is doing.

Fixed-wing drone

In this section we'll build an FPV fixed-wing drone with a fully featured autopilot that'll enable it to fly entirely on its own, from take-off to landing. A downwards-facing camera can also be added to this platform to turn it into a mapping drone at a later stage.

Choosing the components

We primarily want to create an FPV aircraft, so the only gear it'll need to carry is a small FPV camera and transmitter. Consequently there's no significant payload to be carried. However, since it would be nice to be able to turn it into a mapping platform it would be useful to have the option of carrying a small digital camera.

Estimating the weight of components

The first thing to do is make some educated guesses about the weights of the gear we'd like to use, since this information will be needed to help refine our choice of airframe and motors. As with most drone builds, searching the Internet for inspiration and guidance is a good place to start. This can also provide useful ideas about suitable aircraft platforms that you could use.

The hardest thing to estimate is the weight of the airframe and servos. The best way to get a weight is to search the Internet for R/C aircraft designed to carry similar equipment, such as a mapping camera. This will give you a feel for the size and weight of aircraft that'll be required. The estimated weights of the equipment on this drone are:

- Autopilot system – 150g
- FPV camera and transmitter – 100g
- Motors/ESCs – 150g
- Empty weight of airframe including servos – 800g
- Battery – 500g
- Optional camera for mapping – 200g

The approximate weight of this drone will therefore be about 1.9kg, so say 2kg.

Choosing a motor/ESC and propeller

Now that we have a rough guess of the drone's total weight we need to find a motor that's powerful enough to lift it. As you'll (hopefully) remember, a drone that doesn't need to perform acrobatics should have a power-to-weight ratio of about 80W/lb. Converting 2kg to pounds will give a weight of 4.4lb. The product of 4.4lb and 80W/lb means that we need a motor and prop combination capable of providing a total of about 350W.

After going through the thrust data tables for various motoros, I've chosen a generic 2814 1,100kV motor that's capable of producing about 352W with a 9 x 6in propeller on a three-cell LiPo. Using a four-cell LiPo allows the motor to produce over 500W with the same propeller. This motor is more efficient in terms of the g/W ratio using a 3S battery, so we'll go with that set-up. However, if you decide to have a more sporty drone you can easily upgrade to a four-cell LiPo.

Looking at the thrust table the motor draws a maximum of 29.4A with the 9in propeller on a three-cell battery, so using a 40A ESC will be a safe option.

Selecting a suitable aircraft

The next step is to find a suitable aircraft to use. I generally prefer to use flying-wing style aircraft, as they're easier to store and transport to and from the flying field. Also, since flying wings only have two control surfaces and therefore two servos there are fewer things to plug in and configure.

When looking at an aircraft to buy make sure that it has enough payload space to fit all of your electronics, such as autopilot and GPS. The other factor to look for is to ensure that it can carry the weight of your equipment. The easiest way to do this is to simply read the specifications of various aircraft to see their maximum take-off weight. But if this information isn't available you'll need to do some maths to get a feel for the size of aircraft you're going to need.

To do this we'll start by looking at the wing loading equation to see what size wing is required to carry our equipment. As you might recall, we want a wing loading of about 20oz per square foot, so that by putting this information into the equation we can find the approximate wing area that we need:

$$\text{Wing loading} = \text{weight (pounds)} \times 2{,}304 \div \text{wing area (square inches)}$$

$$\text{Wing area (square inches)} = 4.4 \times 2{,}304 \div 20 = 506.8\text{in}^2$$

ABOVE The completed Skywalker X6 drone during its first test flight. As a precaution against any initial problems, it's a good idea to make the first test flight in a large open area, so you have plenty of space to perform an emergency landing if necessary. *(Author)*

So if we assume our wing chord is about 8in wide (about 20cm), the wingspan of our drone will need to be about 63in (506.8/8), or 1.6m. Knowing the approximate wing area means we can now look around for drone platforms that are big enough.

If you've never built an R/C aircraft before it would be a good idea to buy a bind-and-fly (often abbreviated to BNF) drone with enough capacity to carry your camera and FPV gear. Bind and fly aircraft arrive fully built, with all the basic components – such as servos, motors and propellers – pre-installed. All you need to do is add the autopilot and R/C receiver in order to convert it to a drone.

I've decided to go with the Skywalker X6 aircraft, as it's a flying-wing design with a large central fuselage that can carry all the gear we need. It's also specially designed for FPV applications. The total take-off weight for this platform is around 2kg, which fits within our weight limits.

Refining the weight estimation to choose a battery

Now that we've chosen all the initial items we need we can go over the weight estimation again to arrive at a more accurate figure. As it turns out the total weight of the Skywalker X6 is just 200g lighter than estimated, so we can increase our battery weight from 500g to 700g and yet still have the same all-up weight. After looking around I decided on a 5,000mAh battery, although the total weight allowance would actually allow us to use a slightly higher capacity.

Since we're using a 40A ESC with our motor we must ensure that the battery we choose is capable of providing enough current. The proposed battery is a 5,000mAh 20C three-cell LiPo, so the maximum continuous discharge is 5Ah x 20C = 100A, which is more than enough for our motor. We could also get away with a lower discharge battery if we wanted to save some weight.

Estimating the flight time

Estimating the flight time of this drone before we build it can be quite difficult, since we don't know how much throttle we'll need in order to fly. This will often require a test flight to measure the current flow during normal flying conditions. However, if we assume we need to fly at half throttle, looking at the motor specs/thrust table we find the motor will produce about 112W, while our battery can provide 55W of power. Putting these figures into the flight-time equation will give:

$$55.5 \text{ (battery power)} \div 112 \text{ (motor power)} \times 60 = 29 \text{ (minutes)}$$

This assumes that we're able to cruise at half throttle, but it nevertheless provides a useful rough guess of how long we can fly.

Building the aircraft

The first steps of building an R/C drone are just like any other R/C aircraft. Once the aircraft itself is built we'll proceed with integrating the autopilot and other systems.

Building the wings

1 Some R/C aircraft arrive as a collection of foam pieces that you have to put together. Often there'll also be an instruction manual that will tell you the basics in terms of building the frame and attaching all the parts. If the product doesn't include a manual or you get stuck, the best thing to do is go online and ask someone in the R/C community (try www.dronetrest.com) for help.

Because foam isn't the strongest of materials most foam wings come with carbon rods that serve as wing spars, to increase their rigidity. These need to be glued into the wings. However, if your aircraft has wings that can be removed you don't want to glue the main spar. So with this build we need to glue the smaller top spars on to the wing but don't want to glue the main, larger, wing spar.

2 We'll also need to glue the servo horn connectors to the wings, to enable the control surfaces to move. Be sure to use a foam-friendly glue, as some will actually eat away the foam!

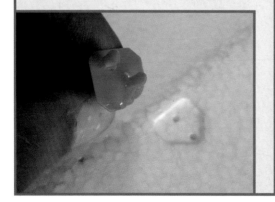
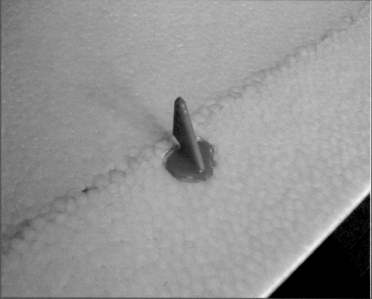

3 Most foam wings don't include any hinge mechanism for the control surfaces so you should always reinforce them with some packing tape or similar. In this case I've used fibreglass-reinforced tape, as it's much stronger. When placing the tape on the hinge ensure that it won't prevent the control surface from moving up and down, otherwise your aircraft won't be very controllable. Also, take your time to ensure there are no creases in the tape as these would make the wing less efficient due to aerodynamic drag.

4 Now is a good time to attach the servos to the wings. Most aircraft will have specific cut-outs in the foam where the servos should be mounted. Since the Skywalker X6 is a flying-wing design we only need to use two servos.

The servos are also simply glued into position. Ensure that the servo arms are aligned to the orange servo horns you've already glued into position on the control surface. However, before gluing the servos down you should apply some 5V power to them to ensure they're reset to their neutral position. The easiest way to do this is to use a servo tester, but you can also connect the servo directly to your receiver and power this with an ESC connected to a battery.

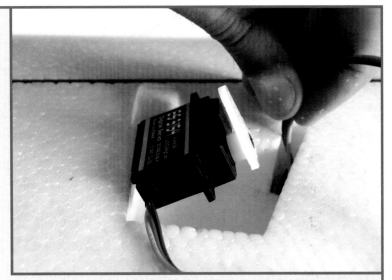

5 Next we need to connect the pushrod between the servo and the wing control surface. This enables the servo to move the wing surface up and down to control the aircraft. Consequently it's very important that it's fastened securely, as it's the critical link that makes your aircraft controllable in the air. Using some pliers, bend round the end connected to the servo so that it won't come undone.

It's generally better to mount your servos and control linkage on the top of the wing, as this protects them during landing by preventing anything from getting caught in the grass. However, it's actually slightly more aerodynamically efficient to have the control linkages on the bottom of the wing.

If you don't like the plain while look of your aircraft, you can always use some foam-friendly spray-paint to add character to your drone. It's best to paint your drone before assembly and to use bright colours to ensure your aircraft is easily visible. Some pilots might choose to use coloured parcel tape – while it is heavier than paint, it also adds extra protection to the foam.

6 Though the Skywalker X6 has a cool feature that allows the wings to be detached for storage and transport without any wires getting in the way, this inevitably adds complexity to the build, as we need to solder d-sub (d-subminiature) connectors to the servos and other equipment mounted on the wings. The Skywalker also allows us to mount extra gear on the wings such as our FPV transmitter or R/C receiver, to ensure they get the best possible reception away from other electronics. However, this feature is intended more for long-range FPV systems. Since we're only going to be flying within visual line of sight (as per UK regulations) it won't be needed, as it's much easier to mount all our equipment inside the main body of the aircraft.

7 Using a standard servo extension cable, cut it in half and solder a d-sub connector in between. Since a servo cable uses three wires we'll solder one to each of three connectors at the back of the d-sub. It doesn't matter which three pins you choose just so long as you're consistent at both ends of the connector, so that the signal can be routed correctly from one end to the other. If you plan to mount your equipment to the wing you should solder the extra connectors at this point.

Once the soldering is complete we can attach the connectors to the plastic wing connectors.

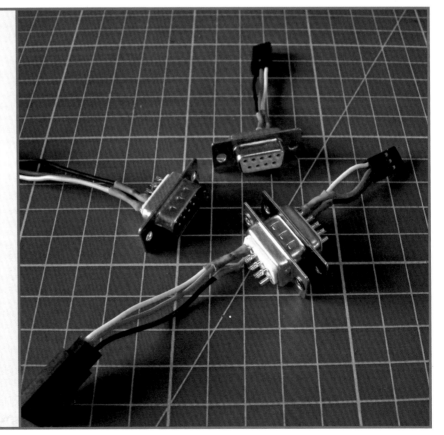

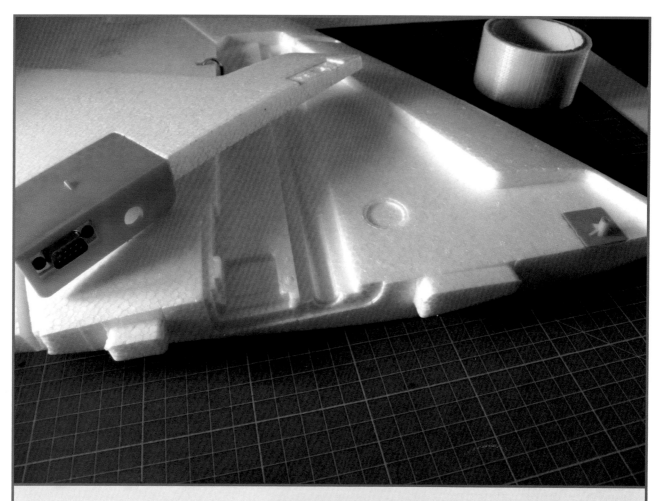

8 Next we have to glue the wing connectors to the underside of the wing. There's a large foam cut-out that we need to attach on the bottom of the wing, as the main spar fits between this and the wing. However, before we secure it in place we must first link our servo to the wing connector and secure the wires neatly using some tape.

9 Since we might want to upgrade or add extra equipment to the wings in the future it's best not to glue the foam cut-out to the wing permanently. This will also allow access to the servo if you ever need to replace it. Just use sufficiently strong fibreglass tape – and enough of it – to ensure that it won't come undone. Try to avoid introducing creases while you apply the tape.

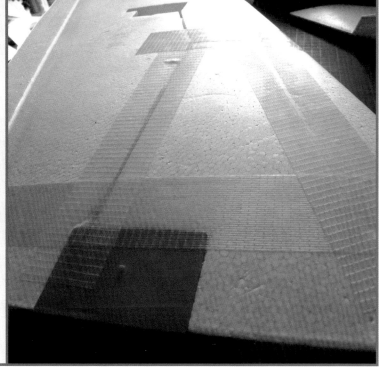

10 The final step is to secure the winglets to the wings. These help to keep the aircraft flying straight. The winglets on this airframe are simply glued to the edges of the wings. In addition the wingtips should be reinforced with extra tape, as they usually hit the ground during landings.

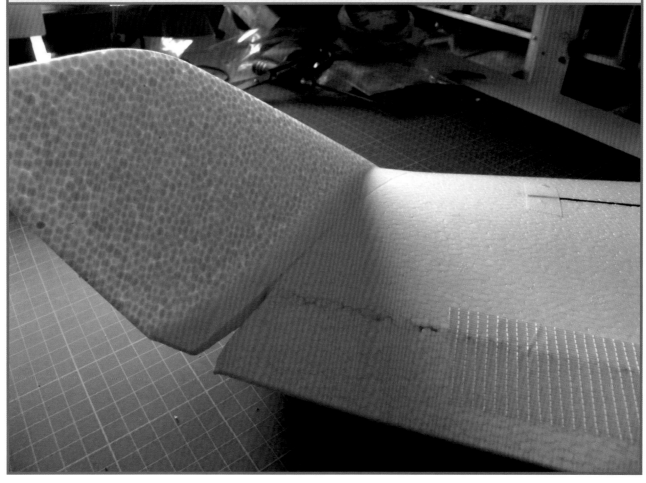

Mounting the motor

11 Because foam is quite weak we can't screw the motor into it, nor glue the motor directly to it. Instead it has to be attached to the plastic motor mount that's supplied, which is then glued at the back of the aircraft since it's a pusher configuration. Other designs might include an aluminium motor mount bracket which is glued to the foam. Using plywood motor mounts sandwiched between the foam is also fairly common.

Soldering the ESC and power module

12 Before we can build the main body of the aircraft we should first install and solder primary electronics like the ESC, motor and power module inside it. The motor is connected to the ESC via bullet connectors, so we don't need to solder them together directly. However, due to the size of the aircraft we will need to solder some extension wires between the battery and the power module so that the battery (mounted towards the front of the aircraft) will reach the power module (mounted near the middle).

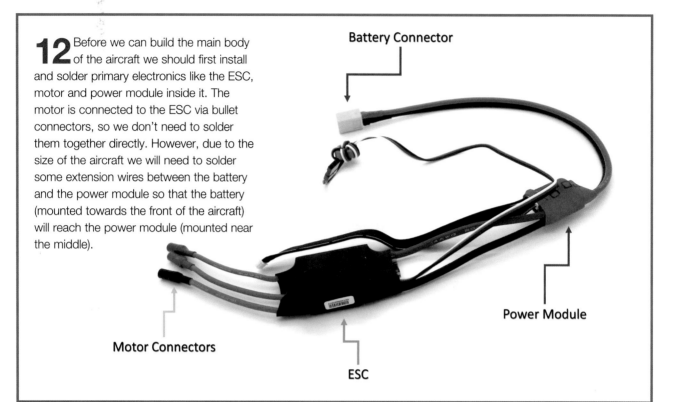

Battery Connector

Power Module

ESC

Motor Connectors

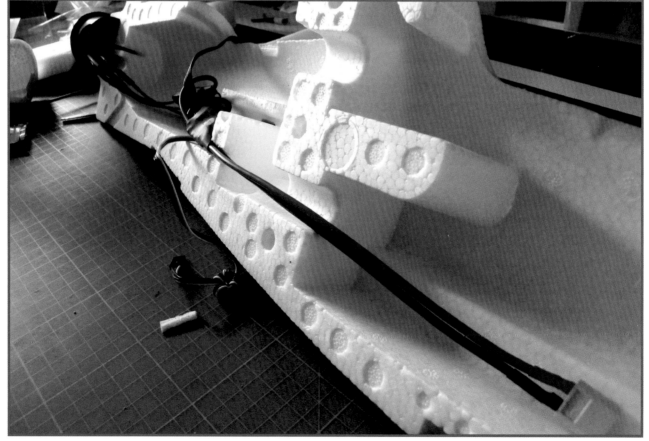

13 You can also connect the motor and ESC bullet connectors, but make sure you do this the correct way round so that the propeller spins in the correct direction. The direction will depend on the propeller you're using, but the motor and propeller should spin to blow air backwards, which propels the aircraft forwards. To test this we can make use of the servo tester again, or you can connect the ESC to the throttle channel on your R/C receiver. It's best to do this now, as it can be fiddly trying to access these parts once the fuselage is glued together.

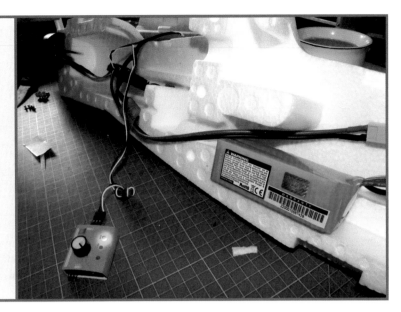

Adding your FPV equipment

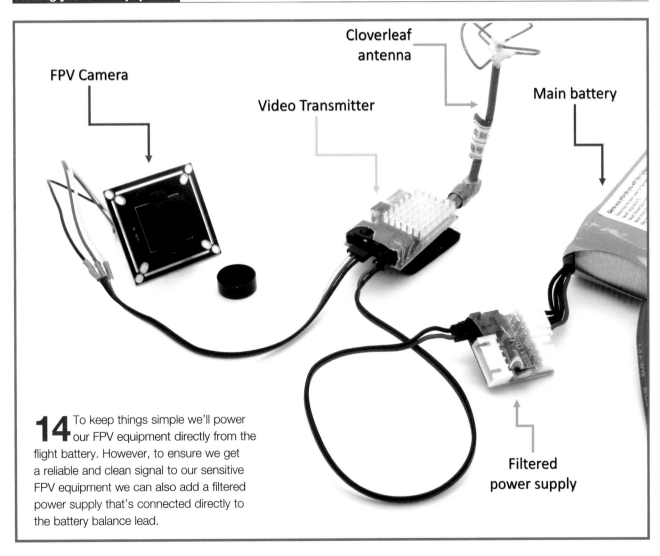

14 To keep things simple we'll power our FPV equipment directly from the flight battery. However, to ensure we get a reliable and clean signal to our sensitive FPV equipment we can also add a filtered power supply that's connected directly to the battery balance lead.

15 You should always perform a quick test of your equipment before mounting it. Simply connect the FPV gear and turn it on by connecting a battery to ensure it's working correctly. It's also an easy way to see which way round the camera should face so that we don't accidentally mount it upside down!

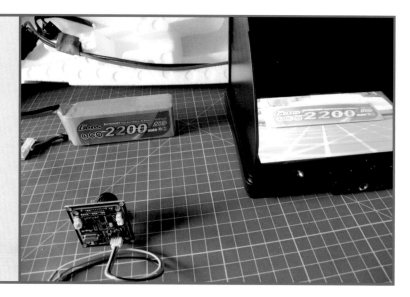

16 Before you glue the two halves of the fuselage together you should take a second to consider what other equipment you need to mount. Since we want our camera to stick out the front (so that we can see where we're flying) we'll also need to cut a hole for the camera lens to see through.

Assembling the main body

17 To finish the assembly of the main body we simply need to glue the two halves together. When gluing make sure that everything is lined up properly. You can use tape to hold the frame together while the glue is drying.

18 Once the frame has been glued together you should also add some parcel or fibreglass tape along the belly of the aircraft to help protect the foam during landing. Use some double-sided foam tape or velcro to secure the FPV gear in the front payload compartment. The battery mount should also be glued in place.

Installing the autopilot

19 On this drone we're using the APM2.6 autopilot system running ArduPlane firmware, which is part of the same family as the ArduCopter firmware we used in the aerial photography quadcopter build. The same ground control station software is therefore used to set up and configure the autopilot. In terms of the exact way to connect your autopilot you'll need to refer to its instructions, as this will be specific for each autopilot system.

Although not essential, it's always best to mount the autopilot as close to an aircraft's centre of gravity as possible. In this aircraft there's a suitable payload compartment in that exact location, in which the autopilot board and receiver should be secured with double-sided foam. Using double-sided foam has the added advantage of providing a degree of vibration isolation. Since the receiver is something you'll commonly use on other aircraft too it's best to mount this with Velcro to make it easier to remove.

20 There's no more room left inside the rear compartment, as the space underneath the autopilot is reserved for a mapping camera to be mounted at a later stage. The telemetry module is therefore mounted to the underside of the rear compartment's lid, through which a hole is punched so that the module's antenna can stick out the top.

21 With both payload compartments open you can see where all the equipment is mounted. The GPS module is in the front compartment, away from the ESC and other electronics that generate lots of interference.

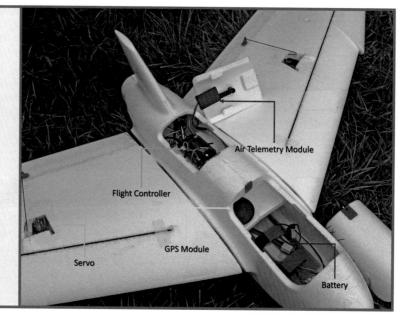

Checking the centre of gravity

22 When attaching the battery you should move it forwards and backwards to ensure the aircraft's centre of gravity is in the right place. Its exact location will be included in the aircraft's instructions. In the case of the Skywalker X6 it should be located between the main spar and the d-sub connector. By holding the fully assembled aircraft at that point you'll be able to see if its centre of gravity has ended up where it should be. If not, you must slide the battery forwards or backwards until you get it balanced. In extreme situations where moving the battery doesn't balance it you might need to move other equipment around too, or add extra weight at the front or back. However, it's very important that you get the centre of gravity just right or your aircraft could become unstable – which could result in a crash.

Finishing up

23 Now that the drone assembly is complete, the last thing to do is to connect your autopilot to your computer and set it up. Settings will involve selecting the vehicle type so that the autopilot knows how many servos you're using. You'll also need to calibrate your ESC and R/C transmitter and check that the servos are moving correctly. However, all the guidance for this will be included in the autopilot's documentation. If you ever have any issues or need help you can also contact me, or anyone else in our community forum, at www.dronetrest.com.

Once again, ensure that during set-up the propeller is removed to avoid inflicting damage or injuries if the motor should suddenly turn on.

BELOW The completed Skywalker X6 drone awaiting its first flight. The propeller hasn't yet been attached for safety reasons, as the autopilot has yet to be configured.

CHAPTER 6
FLYING AND SAFETY

(Author)

**OPENING SPREAD
When launching your aircraft make sure that you throw it at a very slight upwards angle that isn't too steep; otherwise you might cause it to stall.**

Now that you have an idea of how drones work and how to build one we need to discuss flying them safely, because although some drones essentially fly themselves, you should still be a competent radio-control pilot so that you can take over if something doesn't work as it should.

The regulations and safety aspects that I'll be talking about here are, inevitably, subject to change. Although you can probably use the information presented in this book as a guide for flying your drone legally in the UK, it's important that you always take the time and effort to ensure that you're flying in a safe manner and don't break any local regulations.

Safety information

Compared to other countries the UK currently has fairly relaxed and fair regulations when it comes to flying drones for hobby use, so as long as we all keep flying safely and don't do anything dangerous – like flying a drone over a major airport – everyone will be able to continue enjoying this hobby without the need for stringent regulations. However, if you plan to use your drone for commercial purposes you'll need to obtain a certificate permitting you to do so. This involves going on a training course and completing an examination to ensure you know how to fly and operate your drone in a safe and competent manner.

The legal parts

The CAA have addressed many of the legal issues of using model aircraft and drones in various Air Navigation Orders. These are rules published to ensure the safety of the national airspace, so breaking these rules could result in criminal prosecution.

Below are a few relevant excerpts from the CAA regulations.

The first articles, 137 and 138, are generalised towards model aircraft but are also applicable to drones (which are essentially advanced model aircraft):

Article 137 – 'A person must not recklessly or negligently act in a manner likely to endanger an aircraft, or any person in an aircraft.'

Article 138 – 'A person must not recklessly or negligently cause or permit an aircraft to endanger any person or property.'

Now for some specific articles relating to the flying of drones:

Always remember:

You are responsible for each flight	You are legally responsible for the safe conduct of each flight. Take time to understand the rules - failure to comply could lead to a **criminal prosecution.**	**Keep your distance**	It is illegal to fly your unmanned aircraft over a congested area (streets, towns and cities). Also, stay well clear of airports and airfields.
BEFORE each flight, check drone for damage	Before each flight check that your unmanned aircraft is not damaged, and that all components are working in accordance with the **Supplier's User Manual.**	**Keep your distance 50 metres**	Don't fly your unmanned aircraft within 50m of a person, vehicle, building or structure, or overhead groups of people at any height.
Drone is in sight at all times	You must keep the unmanned aircraft within your sight at all times.	**Consider rights of privacy**	Think about what you do with any images you obtain as you may breach privacy laws. Details are available from the Information Commissioner's Office.
**YOU are responsible for avoiding collisions**	You are responsible for avoiding collisions with other people or objects - including aircraft. Do not fly your unmanned aircraft in any way that could endanger people or property.	**Permission to use drones for paid work**	If you intend to use an unmanned aircraft for any kind of commercial activity, you must get a 'Permission' from the Civil Aviation Authority, or you could face prosecution. For more details, visit www.caa.co.uk/uas

RIGHT CAA guidelines for flying your drone.

Article 166 (Small Unmanned Aircraft)
- [1] 'A person shall not cause or permit any article or animal (whether or not attached to a parachute) to be dropped from a small aircraft so as to endanger persons or property.'
- [2] 'The person in charge of a small unmanned aircraft may only fly the aircraft if reasonably satisfied that the flight can safely be made.'
- [3] 'The person in charge of a small unmanned aircraft must maintain direct, unaided visual contact with the aircraft sufficient to monitor its flight path in relation to other aircraft, persons, vehicles, vessels and structures for the purpose of avoiding collisions.'
- [5] 'The person in charge of a small unmanned aircraft must not fly such an aircraft for the purposes of aerial work except in accordance with a permission granted by the CAA.'

If you plan to use your drone for filming or photography, there are some additional regulations published in –

Article 167
- [1] 'The person in charge of a small unmanned surveillance aircraft must not fly the aircraft in any of the circumstances described in paragraph [2] except in accordance with a permission issued by the CAA.'
- [2] 'The circumstances referred to in paragraph [1] are:
  a) over or within 150 metres of any congested area;
  b) over or within 150 metres of an organised open-air assembly of more than 1,000 persons;
  c) within 50 metres of any vessel, vehicle or structure which is not under the control of the person in charge of the aircraft; or
  d) subject to paragraphs (3) and (4), within 50 metres of any person.'
- [3] 'Subject to paragraph [4], during take-off or landing, a small unmanned surveillance aircraft must not be flown within 30 metres of any person.'
- [4] 'Paragraphs (2)(d) and (3) do not apply to the person in charge of the small unmanned surveillance aircraft or a person under the control of the person in charge of the aircraft.'
- [5] 'In this article "a small unmanned surveillance aircraft" means a small unmanned aircraft which is equipped to undertake any form of surveillance or data acquisition.'

In summary, here are the basic takeaways to ensure safe and lawful drone operation for hobby use in the UK:

- The operator must not endanger anyone or anything by being unfamiliar with the current laws and regulations.
- Do not endanger anyone or anything.
- Make sure that your flying location is appropriate and safe.
- Ensure that you maintain a visual line of sight with your drone at all times. If you're flying via FPV you'll need a second person to do this for you, although this person doesn't need to know how to fly.
- If you're making money with your drone, you're using it for aerial work and need appropriate qualifications to do so legally.
- Do not constitute a nuisance.
- Do not invade the personal privacy of anyone.
- It's a good idea to have appropriate liability insurance to protect you in the event of an incident that leads to a claim against you. Joining an organisation such as the BMFA (British Model Flying Association) will provide this as part of their membership.

Where can I fly?

Apart from what's been mentioned above about not flying close to people or in built-up areas, there are several tools available online which are regularly updated with information about where you can and can't fly your drone. In general you can't fly close to any active airports and from time to time the CAA will issue a NOTAM (notice to all airmen) about temporary flight restrictions in certain areas. A relatively nice application that you can use on your android device is called NoFlyDrones, which includes a map showing the areas where you can't fly your drone. Alternatively, if you don't want to buy the app

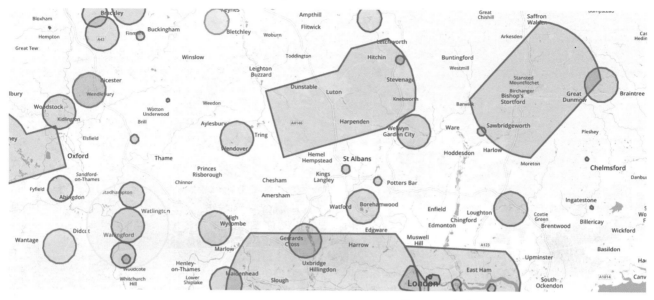

ABOVE Screenshot of various restricted airspace zones where you can't fly a drone. *(Author)*

BELOW Using a simulator to which you can connect your R/C radio provides a great way to practise your flying skills without the risk of crashing your drone. *(Author)*

you can view it on your browser by going to this link: http://bit.ly/noflyzones.

In general you can fly your electric drone in most public spaces such as parks, assuming there are no by-laws restricting or prohibiting the use of model aircraft. If you're flying on private land it's important that you first obtain permission from the landowner. No matter where you fly make sure that it isn't a crowded space. You must also make sure that you're competent in your piloting abilities when flying in areas with less space, so as to avoid crashing.

Learning to fly

Thanks to the built-in flight controller, flying a drone is much easier than regular R/C aircraft or helicopters.

Start with the basics

When people build their first drone they often want to add all the bells and whistles available. However, my suggestion if you're just starting out in the hobby is to begin with the barest minimum of equipment when you make your first few flights, such as just the autopilot. Once you're happy your aircraft is flying and that you're comfortable flying it, then you can start adding extra equipment like a camera or FPV gear. The reason for starting this way is that you're risking less of your investment until you're more confident in your skills. Also, adding more components makes the overall system more complex, which increases the risk of making a silly mistake. By building everything up step by step you can ensure that it's all set up correctly.

Simulators

There are several R/C flight simulators available on the market, including some free ones. This is a great way to get comfortable flying R/C aircraft without having to care about crashing – not to mention that it's also great fun! Most R/C flight simulators include a USB cable that you can connect with your R/C radio to control your aircraft, but when choosing an R/C flight simulator it's a good idea to check if your radio is compatible. If not, you can also buy cheap dedicated simulator R/C radio controls on the Internet.

Many R/C flight simulators also include flight tutorials and challenges designed for all skill levels. This is a great way to learn to fly,

ABOVE **A cheap miniature quadcopter toy is a great way to practise your flying skills.** *(Author)*

starting with the basics all the way to advanced acrobatic flying. Learning on a flight simulator will greatly improve your flying skills as well as give you the confidence to try new things without worrying about breaking your aircraft. Modern R/C flight simulators have very realistic flight physics, so if you're flying an aircraft that's already included in the game the chances are that when you fly it in real life you'll hardly be able to tell the difference.

In my opinion one thing that R/C flight simulators don't get entirely accurate is the wind. Although many do include advanced weather and wind options it never feels the same as in real life; so when you first start flying make sure that you wait for a calm day, as it'll make your first flights much easier.

As discussed previously, some autopilot systems include a simulator feature. Often this is also a great way to become familiar with how your autopilot works, as you can plan missions, test automatic take-off and landing features and so on, all from the comfort of your PC.

Cheap toy quadcopters

Flying fixed-wing aircraft is very different to multirotors, and I could even argue that multirotors are actually easier to fly. If the idea of using a flight simulator doesn't appeal to you the other great option is to buy a cheap mini quadcopter toy. These are often very basic but will still teach you the fundamentals of how to fly and hover a multirotor, and they often cost less than £20 for the entire kit. If you can fly one of these toy quadcopters well, then flying a fully featured GPS multirotor drone will be no problem for you, as the extra sensors and advanced features of the larger drones make them easier to fly.

Pre-flight safety checks

Let's face it, unless you're going to be using your drone for commercial use you're unlikely to go through a detailed checklist step by step before each flight (although it's never a bad idea, especially if you're using expensive gear on your drone). So in this section I'll only provide pointers of the key things you should go over before each flight. This way if you do want to make a pre-flight checklist you can tailor it to suit your own drone. Also, although some of the things mentioned here should be checked before each flight you'll only need to check others when you add or remove equipment.

Check your batteries

Obvious things to check include making sure your batteries are properly charged on both your drone and your ground station tablet/laptop, video monitor and, most importantly, your radio-control device. There's nothing more annoying than arriving at your flying location only to realise one of your batteries wasn't fully charged.

Checking the environment

Before you go out and fly it's also a very good idea to check if you're allowed to fly in that location, and there are no temporary restrictions. Once at your flying spot make sure it isn't too crowded, as you're not allowed to fly within 50m of other people or buildings. Throughout your flight, if you see or hear any low-flying manned aircraft it's a good idea to keep well away and even land if you're unsure. Lastly you should check the weather forecast to ensure it isn't too windy.

Checking your drone

A very useful check is to make sure that all your equipment is securely fastened and won't come loose during flight, especially the flight battery. I remember when I made this mistake with a plane where I hadn't secured the battery properly, and during a tight turn the G-force caused it to fly out of the battery compartment, leaving my plane with no power or control. All I could do was sit there as my plane spiralled down to the ground… It doesn't matter if you have the very best safety features on your drone (such as a parachute), if the entire platform has no power nothing will work!

Although it's something you'd address during your drone build, it's still a worthwhile precaution to make sure that all your cables are securely attached and won't get in the way of your motors or propellers. Also, ensure that there are no chips or cracks in your propellers, as you don't want extra vibrations or, worse, a propeller breaking mid-flight.

Once you've made sure everything is secured, the next step is to check that your autopilot system is working properly. Since all flight controllers initialise the accelerometers and gyros when you turn them on it's important to make sure your drone isn't moving and is on a level surface when you do so, to ensure everything initialises properly. If you're using a GPS receiver with your drone you'll need to check that it's locked to enough satellites to achieve a good position estimate and a home location before you take off. The way you verify these will vary from system to system, but they can usually be verified via a status LED, or the ground control station app on your tablet, smartphone or laptop. Since there are many different autopilot systems it's hard to give a definitive list of things to check, but the documentation should tell you what the most important things are. Some flight controllers also have built-in safety features to prevent you from taking off if they think there's a problem.

The final thing to check is the mechanical aspects of your drone. In the case of a fixed-wing drone you should move the sticks on your radio control to make sure the plane's control surfaces are moving as they should. If you have a gimbal, make sure that it has initialised properly and is level (as it's very annoying to get home and find all the footage you took was tilted to one side!). Also, you should give all the screws a once over from time to time, as vibrations caused by the motors can cause them to loosen.

BELOW Make sure all the systems of your drone are working correctly before each flight. *(Author)*

Things to check before your first flight

Once you're ready to take your drone out for its first flight there are further things you should check to avoid some common pitfalls.

R/C gear

Always check that your R/C gear is communicating correctly with your autopilot. Most autopilot systems enable you to check this via the R/C calibration process that's part of the autopilot's set-up procedure. But you must make sure that no mixing has been set up on your radio. An easy way to check is to simply move a single stick on your R/C radio – only one channel should respond on your flight controller. You must also make sure that the channels are correct, so that if you move the throttle stick from low to high you should see the same response on the flight controller configuration screen. If another channel moves then something isn't connected correctly. Also, if you move your stick from low to high and the flight controller reads the input as high to low, then you'll need to reverse that channel, either on the radio or preferably on the actual flight controller software.

Autopilot orientation

Most autopilots have clear arrows showing which direction is forward. However, they also have the option to change the orientation in the software. An easy way to check this is to connect your drone to your PC, tilt the drone upwards and check that it responds as it should on the HUD (*ie* it should show that your drone is pitching upwards).

Flight modes

It's critical to set up your flight modes before your first flight. Make sure that you have the default flight mode as 'stabilise', and check that the mode changes correctly when you change the flight mode switch on your R/C radio.

Check the connections

A critical step during construction is to ensure that all the equipment is connected in the correct way. When something isn't working as it should the most common causes are either a loose connection or something being

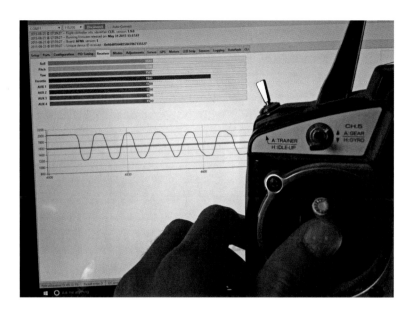

ABOVE Your flight controller software enables you to check that it's reading your R/C inputs correctly. In this image the throttle channel stick is pushed all the way up, which corresponds to the red bar on the screen that represents the throttle input. If you move the throttle stick up, and the red bar moves down, you'll need to reverse the channel. This process is very similar for all flight controllers but the software might look slightly different. *(Author)*

connected in the wrong place. So it's worth taking a few extra minutes during your drone build to double-check everything. Before a flight you can also connect up your autopilot and check the systems are working correctly via the ground control app before take-off.

Servos

If your aircraft uses servos (fixed-wing or tricopter) you must ensure that these are reacting the way they should, as it's not uncommon for them to react in reversed order. If you pitch down on your R/C gear you should see the servos cause the drone's elevators to tilt downwards. If they don't you'll need to reverse the servos in the flight controller settings. Similarly, when your aircraft is in stabilise mode and you tilt the entire drone downwards you should see the elevators tilt upwards to try to keep the aircraft stable. You should always make sure the servos are reacting in the correct directions before you try and fly a drone.

Checking ESC calibration

Before your first flight you must also make sure that your ESCs all have the same calibration.

To check this, remove the propellers from the motors and turn your drone on. Then slowly apply throttle. If all the motors start at the same time then the ESC calibration is correct. If some start before others you'll need to recalibrate your ESCs as discussed in earlier chapters.

Motor/propeller direction

A critical step on a multirotor is to ensure that your motors/ESCs are connected to the correct outputs on the ESC and that they're spinning in the correct direction according to the flight controller documentation. If any of them are spinning in the wrong direction your multirotor will flip down into the ground on take-off. Also make sure the propellers are connected the right way around. It might sound kind of obvious, but it happens more often than you'd expect!

Centre of gravity

Much more important for fixed-wing drones is to ensure that their centre of gravity is correct, as this will ensure that they fly in a stable manner. Having a centre of gravity that's too far forward will make your aircraft nose-heavy, rendering it far more likely to dive into the ground when you take off. Having the centre of gravity too far back can cause your aircraft to become unusable during flight, particularly if it stalls, resulting in a crash. The instructions included with your aircraft frame will usually tell you where the centre of gravity is. If not you should ask an expert for suggestions for your particular drone.

For multirotors you ideally want the centre of gravity to be in the centre of the drone. However, if it's slightly off-centre that's usually not a major issue, as the motors are able to cope.

Your first flight

Your first flight can be a scary experience, especially if you've dedicated a lot of time to building your drone. However, in order to make the process less likely to end in disaster it's best to practise on a flight simulator before flying in real life. If you know anyone who's already an R/C pilot it's also a good idea to ask them for guidance. In addition some autopilots include safety features designed to help prevent you from crashing. Assuming you've followed the instructions for your flight controller and gone through the proper set-up process and pre-flight checks your first flight should go smoothly, and your autopilot will take care of the hard parts for you.

When flying your drone for the first time make sure you're in a large open area away from other people, and wait for a calm day. Also, don't try your first flight in fully automatic mode (where the drone controls the take-off and landing for you), as you first need to verify that it's flying properly.

During your maiden flight you should keep a relatively low altitude and settle for making sure that the drone is behaving the way it should – keeping level and responding correctly. If it doesn't react properly and is very sensitive or wobbly you might need to land again to fine-tune the PID values, as discussed in Chapter 4.

Once you're happy your drone is behaving as it should, the next step is to progress to more advanced functions such as 'altitude hold' and 'position hold'. When you're satisfied that the basic functions are all working properly and you have gained confidence in flying your drone you can move on to trying something more complicated such as planning an automatic mission.

Buddy box

If you know anyone who's a decent radio-control pilot chances are that they'll be happy to help teach you to fly. A very useful feature found on almost all R/C radio controls is the 'buddy box' feature used for pilot training. This enables two radio control units to connect to each other (wirelessly or via cable), with one unit being the master (controlled by the trainer) and the other being a slave (controlled by the trainee). This way if something goes wrong while you're learning to fly the pilot with the master R/C can take over control at any time to prevent a crash.

Flight aids/geofence

If you don't have the option of a buddy box the next best thing is to enable some flight aids on your drone. Most GPS autopilots allow you to specify an imaginary area around the take-off location, and if you fly outside of that box the autopilot will take over to prevent you from losing your aircraft. You can think of this geofence as being like a cylinder, with a maximum and minimum altitude above the

ABOVE FPV quadcopter drone waiting to take off. *(Author)*

ground and a radius based on your take-off location. If you fly too high, too low or too far away the autopilot will automatically take over and return the aircraft to the centre of this zone.

Launching a fixed-wing drone

Typically fixed-wing drones are hand-launched. You hold the belly of the aircraft in one hand and throw it into the air, while holding the R/C controller in the other. As soon as you throw the aircraft you must quickly use your R/C transmitter to send the drone any course corrections. If you've never flown before you should really get someone else to throw the plane for you so that you can keep both hands on the controls.

Before throwing a drone you should apply around 50% throttle so that the motors starts to produce some thrust, then gently throw the drone at a very slight upwards angle and apply more power. A common mistake is to throw the drone at too steep an angle, which causes it to stall and fall to the ground. Another common error is to throw the drone way too hard or too softly, both of which also usually end in your aircraft hitting the ground. Assuming you have a well-designed aircraft, applying some throttle will produce just enough power while you throw the drone; then as soon as it's been released you need to apply full power to gain altitude.

Using an autonomous take-off feature on your drone will take care of all of this for you, making your first flight much easier. However, before you try an auto take-off you must be certain that everything on your drone is working as it should, usually by conducting a few test flights first. So your first few take-offs will often be under manual control.

As there's no easy way to hold some flying-wing designs during a launch the throwing style for these is slightly different. Instead of holding the belly of the aircraft you need to grip the front of its wings. Then, with a straight arm, you swing the drone around and release it at approximately level flying altitude. This provides an easy way to grip the plane and prevents you from getting your hand jammed into the propeller at the back of the wing.

Where to get help

If you have any questions related to drones the best place to go is the www.dronetrest.com forum, where you can discuss all things drone with either me or any other member of the DroneTrest community. It's constantly updated with new guides and tutorials so you can always improve your knowledge and skills. If you get stuck or aren't sure about what you need to get for your drone you can ask in the help section. There are also many inspirational build examples from other drone builders all round the world.

APPENDIX
THRUST DATA TABLES

Thrust data tables provide us with motor performance data so that we can decide if a motor will be suitable for our drone. During the production of a new brushless motor, manufacturers test its performance with various propellers and at various voltages and this information is published in the motor's thrust data table (also referred to as its 'motor performance data'). Thrust data tables are the principal guideline that you'll use to select the best motor for your drone.

How to read a thrust table
Depending on manufacturer the layout of thrust data tables may vary, but they usually contain the same basic information arranged in a table with the following individual columns:

Voltage
The motors are tested at various voltages that correspond to the various LiPo batteries you'll use with them, such as 11.1V (3S LiPo), 14.8V (4S LiPo) and so on.

Propeller size
A range of propeller sizes is also tested. You'll find that typically the larger propellers are used with higher voltages. This column gives you an idea of the recommended propeller sizes you should use with the motor.

Throttle
This column shows the throttle percentages applied to the motor. Depending on the motor manufacturer some thrust tables show a complete range of throttle values while others only show 50% and 100% throttle. Some tables don't even show the throttle percentage, just the motor current draw.

Amps (current draw)
The Amps column shows the amount of current the motor uses to produce a given amount of thrust. You use this column to see what the motor current draw is at maximum throttle, enabling you to figure out what rating ESC you'll need with your motor.

Thrust
This column shows how much thrust the motor is generating for the given battery, propeller and throttle percentage. You'll use this column quite a lot when trying to find a motor that can generate enough thrust for your drone to fly. A small note of caution here: because these tests are carried out under static conditions the actual real-world thrust value you'll obtain might be slightly different, due to wind and other environmental factors such as temperature and altitude.

Watts (power)
This column shows us the power of the motor. This is calculated by multiplying the voltage and the current at the given test point. The maximum power rating of the motor is when it's running at 100% throttle.

Efficiency (g/W)
This column is particularly useful when choosing a motor as it lets you know how good it is. The more efficient a motor, the longer you can fly. The efficiency of motors is worked out by dividing the thrust by the power: the higher this number, the more efficient your motor is. You'll often find that larger motors with lower kV ratings have much better efficiency than smaller, high kV motors.

Revolutions per minute (rpm)
This column shows you how fast the motor is spinning.

Temperature
Sometimes manufacturers include motor temperature in a thrust data table, to give you an idea of how warm the motor gets with various voltage and propeller combinations. The harder the motor is working, the hotter it'll get. If it gets too hot during use you could end up damaging the magnets. Depending on the quality of magnets used inside your motors, safe operating temperatures can be as high as 150°C. However, this is rather uncommon for general consumer brushless motors, so basically if your motor gets too hot to touch you're probably pushing it too hard.

How to record a thrust data table

If you can't find the thrust data tables for motors that you already have, or that you've salvaged from another drone, you can always perform some thrust tests yourself and work out your own motor thrust table. This will involve constructing or buying a motor thrust test stand. It's fairly easy to build your own as there are countless plans available on the Internet – it only involves some wood, a kitchen scale and a Watt meter. The Watt meter is connected between your ESC and battery to measure the amount of power and the flow of electricity to the motor. The scale is used to measure the thrust produced by the motor. Using this information with various battery and propeller combinations, you can record the values on a table at various thrust levels. Just make sure that the thrust test set-up is properly secured to avoid it lifting off while testing at maximum thrust!

BELOW A selection of thrust data tables for various brushless motors.

VOLTAGE [V]	PROPELLER SIZE	THROTTLE RANGE	AMPERAGE [A] (LOWER IS BETTER)	POWER INPUT [W]	[hp]	THRUST OUTPUT [g]	[N]	[lb]	RPM [rev/min] (HIGHER IS BETTER)	EFFICIENCY [g/W] (HIGHER IS BETTER)	[lb/hp]
14.8V (4S) 16.8V MAX	15" x 5.0	25.0%	2.6	42	0.05	410	4.02	0.90	2820	11.68	18.22
		37.5%	4.3	63	0.08	680	6.67	1.50	3600	10.73	17.74
		50.0%	6.5	93	0.12	980	9.61	2.16	4280	10.54	17.52
		62.5%	8.0	117	0.16	1180	11.57	2.60	4740	10.09	16.58
		75.0%	10.9	161	0.22	1560	15.30	3.44	5400	9.69	15.93
		87.5%	15.2	225	0.30	2090	20.30	4.52	6120	9.28	14.98
		100.0%	19.8	293	0.39	2650	24.98	5.40	6730	8.85	14.75
	16" x 5.4	25.0%	2.7	39	0.05	470	4.61	1.04	2760	12.05	12.81
		37.5%	4.4	65	0.09	770	7.55	1.70	5420	11.85	19.47
		50.0%	6.8	101	0.14	1130	10.98	2.47	4060	11.09	18.31
		62.5%	9.1	134	0.18	1410	13.83	3.11	4560	10.52	17.30
		75.0%	13.1	194	0.26	1890	18.55	4.08	4780	9.76	16.11
		87.5%	18.1	268	0.36	2430	23.85	5.36	5940	9.07	14.91
		100.0%	24.0	353	0.48	2960	29.03	6.53	6540	8.54	13.73
	17" x 5.8	25.0%	2.7	40	0.05	540	5.20	1.17	2920	13.25	21.78
		37.5%	4.8	70	0.09	870	8.53	1.92	3180	12.43	20.43
		50.0%	7.0	115	0.15	1330	13.04	2.93	3840	11.47	18.85
		62.5%	11.4	168	0.23	1770	17.36	3.90	4500	10.54	17.42
		75.0%	16.6	245	0.33	2330	22.85	5.14	5160	9.51	15.63
		87.5%	23.2	342	0.46	2970	29.13	6.55	5760	8.68	14.28
		100.0%	29.8	441	0.59	3600	35.30	7.94	6340	8.16	13.42
	18" x 6.1	25.0%	2.8	41	0.05	550	5.39	1.21	3180	13.41	22.05
		37.5%	5.1	75	0.10	540	9.22	2.07	2940	12.55	20.60
		50.0%	8.9	141	0.18	1510	14.81	3.33	3720	11.53	18.95
		62.5%	13.7	202	0.27	2140	20.99	4.72	4380	10.59	17.42
		75.0%	20.4	302	0.40	2790	27.36	6.15	4980	9.24	15.19
		87.5%	27.8	411	0.55	3550	34.81	7.83	5600	8.64	14.20
		100.0%	36.2	535	0.72	4150	40.70	9.15	6060	7.76	12.75

Data Collected at 11.1 volts with APC 6x4 Prop					
Throttle Setting	Motor Amps	Input Watts	Prop RPM	Thrust (Grams)	Thrust (Ounces)	Efficiency Grams/W
10%	0.41	4.57	4,099	31.9	1.12	6.98
20%	0.93	10.37	6,074	69.3	2.44	6.68
30%	1.64	18.25	7,622	114.3	4.03	6.26
40%	2.56	28.46	9,075	159.6	5.62	5.61
50%	3.93	43.61	10,676	223.8	7.89	5.13
60%	5.61	62.32	12,183	304.5	10.73	4.89
70%	7.79	86.50	13,543	377.1	13.29	4.36
80%	10.73	119.10	15,138	477.2	16.82	4.01
90%	14.39	159.67	16,597	574.6	20.25	3.60
100%	16.38	181.78	17,285	619.8	21.84	3.41

Item No	Volts (V)	Prop	Throttle	Amps (A)	Watts (W)	Thrust (G)	RPM	Efficiency (G/W)	Operating temperature (°C)
U3 KV700	11.1 (3S)	T-MOTOR 12*4CF	50%	2.5	27.75	350	4000	12.61	40
			65%	4.8	53.28	550	4900	10.32	
			75%	6.6	73.26	700	5500	9.56	
			85%	9.1	101.01	870	6300	8.61	
			100%	11.1	123.21	1000	6600	8.12	
		T-MOTOR 13*4.4CF	50%	2.9	32.19	400	3800	12.43	42
			65%	5.6	62.16	650	4900	10.46	
			75%	7.9	87.69	830	5300	9.47	
			85%	10.5	116.55	1000	6000	8.58	
			100%	12.6	139.86	1100	6400	7.87	
		T-MOTOR 14*4.8CF	50%	4.1	45.51	550	3500	12.09	43
			65%	7.7	85.47	890	4500	10.41	
			75%	10.7	118.77	1060	4900	8.92	
			85%	14.5	160.95	1300	5500	8.08	
			100%	17.3	192.03	1460	5800	7.60	
	14.8 (4S)	T-MOTOR 11*3.7CF	50%	3.2	47.36	460	5300	9.71	43
			65%	6	88.80	710	6500	8.00	
			75%	8.2	121.36	870	7500	7.17	
			85%	11	162.80	1080	8200	6.63	
			100%	13	192.40	1230	8700	6.39	
		T-MOTOR 12*4CF	50%	3.8	56.24	580	5000	10.31	43
			65%	7.4	109.52	880	6300	8.04	
			75%	10.3	152.44	1100	7300	7.22	
			85%	14	207.20	1360	7700	6.56	
			100%	16.8	248.64	1600	8300	6.44	
		T-MOTOR 13*4.4CF	50%	4.7	69.56	730	4900	10.49	47
			65%	9	133.20	1120	6100	8.41	
			75%	12.3	182.04	1400	6800	7.69	
			85%	16	236.80	1600	7400	6.76	
			100%	19.4	287.12	1800	7850	6.27	

Notes: The test condition of temperature is motor surface temperature in 100% throttle while the motor run 10 min.

The voltage (V)	Paddle size	current (A)	thrust (G)	power (W)	efficiency (G/W)	speed (RPM)	Working temperature (°C)
11	EMAX8045	1	110	11	10.0	3650	
		2	200	22	9.1	4740	
		3	270	33	8.2	5540	
		4	330	44	7.5	6200	
		5	390	55	7.1	6700	
		6	440	66	6.7	7150	
		7.1	490	78.1	6.3	7400	36
	EMAX1045	1	130	11	11.8	2940	
		2	220	22	10.0	3860	
		3	290	33	8.8	4400	
		4	370	44	8.4	4940	
		5	430	55	7.8	5340	
		6	480	66	7.3	5720	
		7	540	77	7.0	5980	
		8	590	88	6.7	6170	
		9	640	99	6.5	6410	
		9.6	670	106	6.3	6530	43

Item No.	Volts (V)	Prop	Throttle	Amps (A)	Watts (W)	Thrust (G)	RPM	Efficiency (G/W)	Operating temperature (°C)
MN2212 KV920 V2.0	11.1	T-MOTOR 9545	50%	1.6	17.76	176	4400	9.91	
			65%	3	33.3	289	5500	8.68	
			75%	4.3	47.73	385	6400	8.07	
			85%	5.8	64.38	475	7000	7.38	
			100%	8.1	89.91	617	7800	6.86	

Notes: The test condition of temperature is motor surface temperature in 100% throttle while the motor run 10 min.

Item No.	NO LOAD			ON LOAD			LOAD TYPE
	VOLTAGE V	CURRENT A	SPEED rpm	CURRENT A	Pull g	Power W	Battery/prop
BF3608-15 (460KV)	11.1	0.5	5135	3.6	417	40.0	LiPox3/1047(APC)
				5.4	612	59.9	LiPox3/1238(APC)
				7.6	763	84.4	LiPox3/1365(CF prop)
				5.9	576	65.5	LiPox3/1447(CF prop)
				8.8	930	97.7	LiPox3/1503(CF prop)
	14.8	0.5	6940	5.8	693	85.8	LiPox4/104(APC)
				8.9	1035	131.7	LiPox4/1238(APC)
				11.2	1142	165.8	LiPox4/1365(CF prop)
				8.7	922	128.8	LiPox4/1447(CF prop)
				13.1	1427	193.9	LiPox4/1503(CF prop)
	18.5	0.6	8700	8.1	1015	149.9	LiPox5/1047(APC)
				12.8	1503	236.8	LiPox5/1238(APC)
				15.5	1528	286.6	LiPox5/1365(CF prop)
				12.1	1316	223.9	LiPox5/1447(CF prop)
				17.5	1881	323.8	LiPox5/1503(CF prop)
	22.2	0.6	10335	10.8	1347	239.8	LiPox6/104(APC)
				13.2	1650	293.0	LiPox6/1238(APC)
				18	1750	399.6	LiPox6/1365(CF prop)
				15.8	1752	350.8	LiPox6/1447(CF prop)
				20.3	2103	450.7	LiPox6/1503(CF prop)

LIST OF ABBREVIATIONS

Ah – Ampere hours.
AT – Aerial target.
BEC – Battery eliminating circuit.
BMFA – British Model Flying Association.
BNF – Bind and fly.
CAA – Civil Aviation Authority.
CCD – Charged coupled device.
CMOS – Complementary metal oxide semiconductor.
DIP – Dual in-line package.
DSLR – Digital single-lens reflex.
d-sub – d-subminiature.
DVR – Digital video recorder.
ESC – Electronic speed controller.
FPV – First-person view.
GHz – Gigahertz.
GPS – Global Positioning System.
g/W – Grams divided by Watts.
HUD – Heads-up display.
I-term – Integral term.
IMU – Inertial measurement unit.
IR – Infrared.
ISM – Industrial, scientific and medical.
kHz – Kilohertz.
KV – Measure of the rpm of an engine per volt.
kV – Kilovolt.

LHCP – Left-hand circular polarised.
LiPo – Lithium-polymer.
mAh – Milli-Ampere hours.
mW – Milliwatts.
NiCad – Nickel cadmium.
NiMH – Nickel-metal hydride.
OPTO – Optical isolator.
OSD – On-screen display.
RP-SMA – Reverse polarity sub-miniature version A.
PDB – Power distribution board.
PID – Proportional integral derivative.
PPM – Pulse-position modulation.
PWM – Pulse-width modulation.
R/C – Radio-controlled.
RF – Radio frequency.
RHCP – Right-hand circular polarised.
rpm – Revolutions per minute.
RX – Receiver.
SBEC – Switching BEC.
SD – Secure digital.
SMA – Sub-miniature version A.
TVL – TV line resolution.
TX – Transmitter.
UAV – Unmanned aerial vehicle.
W/lb – Watts per pound.

INDEX

Accelerometers 13, 43-44, 146
Acrobatic flying 48, 57, 145
Advanced airborne radiation monitoring (AARM) 31
Aerial photography and filming 31-32, 34, 76, 85, 90, 92, 98, 143
Aerodrome unguided aircraft 9
Airframes 38-42, 91
   choosing a quadcopter 116
   drilling holes 110
   materials 38
   monocoque 41
   payload area 38, 41, 91, 98, 137-138
Airspeed 47
   indicated 47
Amazon 33
Ambulance drones 34
Antennas 55, 86-88, 138
   attaching 121
   connectors 89
   frequencies 89
   gain 89
   placement 89
   tracking 51-52
Anti-vibration mounts 49, 123
Autopilots 10, 13, 18, 29, 34-35, 38, 42-52, 64, 71, 76-77, 91, 98, 100-101, 107, 124-125, 144-145, 148
   advanced functions 148
   Arduino electronics platform 13
   checking 146-147
   choosing 116
   installing 108-112, 137
   log file 48
   open source systems 12-13
   orientation 147
   setting-up software 114-115, 137, 139
   smartphones 13
Avatar movie 25
Axes of motion 17

Balloons 8
Barometers 44
Batteries 38, 71-74, 91, 124, 127, 150
   attaching and connecting 113-114, 123, 134, 138
   capacity 90, 115
   cell count 71-72
   checking 145
   choosing 94, 100, 117, 125
   discharge rate 94
   flight 135
   LiPo (lithium-polymer) 71-74, 94, 100, 114-116
      charging 73-74
      safety and care 73
      storing 73
      transporting 73
   lithium-ion 71
   NiCad 71
   NiMH 71
   specifications 71-72, 125, 150
Battery connectors 72-73, 119
Battery technology 33
BBC TV 31
Bluetooth 51, 55
Boeing
   B-17 unmanned aircraft 11
   V22 Osprey 25
British Model Flying Association (BMFA) 143
Building a drone 98-139
   choosing components 90-94, 98, 124
   fixed-wing 124-139
      assembling body (fuselage) 136-137
      selecting an aircraft 124
   quadcopter 100-115
   mini FPV quadcopter 116-123
   securing bolts and screws 117
   weight estimating 98-100, 116, 124-125
Cameras, on-board 20-21, 32, 38, 41, 47, 74, 91, 144
   action 76
   downward-facing 17
   DSLR 24, 76
   forward-looking 18
   FPV 75-76, 115-116, 122, 124
   HD video 84, 117, 123
   near-infrared 29
   pan-tilt 35
   professional 25
   SD cards 76, 84
   specialist 33
   thermal 29
   tilting 39
   triggering 76
Catapult launching 9
Centre of gravity 18, 38, 78, 114, 137
   checking 138, 148
Civil Aviation Authority (CAA) 28, 52, 142-143
   Air Navigation Orders 142
   drone flying guidelines 142-143
   NOTAMs 143

CNC machines 39
Commercial use 25, 142-143, 145
   certificate 142-143
Conventional configuration 18
Crashes, and preventing them 19, 48, 83, 138, 144, 148
Cruise missiles 10-11

Dark environment flying 123
De Bothezat, George 10
Definition and classifications of drones 8, 16
Deliveries by drones 33-34
Dronie (selfie photo) 34
Dual operation 32

Electric models 18, 20
Electronic speed controllers (ESCs) 40-41, 43, 49, 59-66, 100-101, 116, 124, 127
   battery eliminating circuit (BEC) 59-60, 110
   cable connection with motor 102, 104, 106, 134-135
   calibration 64-65, 115, 139, 147-148
   choosing 91-92, 99, 116, 124
   common settings 63-64
   connecting to autopilot 11
   connecting to PDB 104, 106, 119
   matching to motor 94
   mounting 105, 118-119
   multirotor firmware 61-62
   output size 59
   programming 62-63
   wiring 105, 107, 118, 120
Emergency service applications 32-33
Empennage (tail unit) 16, 18-19
   elevator 17-18
   rudder 17

Fieseler V-1 'Buzz Bomb' 10-11
First flights 8-9, 147-149
First-person-view (FPV) (video-piloting) 10-11, 16, 18, 34-35, 40, 51, 75-76, 80-90, 143-144
   changing channels and bands 83-84
      DIP switches 83-84
   connecting to battery 135-136
   fitting equipment to a drone 115, 122, 137
   flying with friends 83
   goggles 35, 80, 83-85
   headsets 84
   long-range systems 129
   monitors 83-86
      sunshields 85
   on-screen display (OSD) 90
   receivers 82-85
      diversity 86-87
      DVR function 84
      testing 136
   transmitters 80-84, 115-116, 122, 124
      frequency channels 82-83
      power levels 82
      matching to a receiver 82-83
Fixed wing drones 16-17, 28, 32-33, 38-39, 43, 46-47, 61-62, 92, 124-139, 145, 147-149
   bind and fly aircraft 125
   detachable wings 39
Flight attitude 16
Flight controllers 31, 34-35, 43-44, 60, 91, 115-116, 120-121, 123, 144, 146-147
   configuring to PC 123, 139
   data logging 48
   flight modes 47-48, 147
Flight electronics 119-121, 134-135
Flight time estimating 94-95, 100, 125
Flying bombs 11
Flying control surfaces 16-19
   ailerons 16, 18
   elevons 18
Flying regulations 28, 34, 52, 129, 142-146
   bylaws 144
   choosing the environment 145-146, 148
   NoFlyDrones app 143
   restricted airspace zones 144
   visual line of sight 52, 129, 143
Flying wings 18, 124-125, 149
   Skywalker X6 125, 127, 129, 138-139
Foam aircraft 18, 46, 126
Fukushima nuclear power station disaster 31
Fun flying 90-91
Fuselage (body) 18, 136-137

General inspections 30-31
   sensing 30-31
Gimbals 32-33, 41, 49, 76-80, 91, 113, 146
   balancing 78-79
   controllers 80
   mounting 114
   pitch and yaw adjustment 114
Glider bombs 10
GLONASS network 46
Google 34
Google Earth 30
GPS (global positioning system) 29, 31, 33, 43-46, 49, 77, 108, 111-112, 124, 138, 146, 148
   satellites 47
Ground stations 50-52
   laptops 36, 38, 51
   software 51
   tablets (androids) 50-51
Groundspeed 47
Gyroscopes 13, 43-44, 146

Helicopters – Rotorcraft
History 8-13
Hobby drones 12-13, 25, 34-35, 142
Honeywell RQ16 single copter 25
Hybrid drone (tiltrotor) 25

Indoor flying 47
Insurance 143
Interference 108, 112, 138
    Internal combustion helicopters 20
Iraq invasion 2003 12

Kettering Bug 9-10
Kirby, Simon (Simon K firmware) 61
Kits 16
Korean War 11

Landing 143, 145, 148
Landing gear 41-42
    retractable 42, 111
    return to launch mode 48
Langley, Samuel Pierpont 8-9
Launching (take-off) 142-143, 145, 148-149
Learning to fly 144-145, 148
    simulators 144-145, 148
LEDs 40, 123
Lilienthal, Otto 8
Lockheed DC-130 drone control aircraft 12
Long-range drones 20
Low, Prof Archibald 9
Luftwaffe 10

Manoeuvrability 92
Mapping 17, 28-31, 51, 76, 90-92, 124, 138
    aerial maps 28
    digital elevation models (3D) 29-30
    DroneMapper software 28-30
    geo-referencing 29, 77
    multispectral 29
    NDVI maps 29
Military use 10-11, 25, 28, 33
Motors 20, 57-58, 71, 116, 124
    brushless 57-58
    coaxial arrangement 22-23
    choosing 91, 99
        fixed-wing 92-93, 124-125
            mini quadcopter 116
            multirotor 91-92
    direction of spin 101, 103-104, 118, 148
    in-flight failures 24, 101
    KV rating 58, 61
    naming convention 58
    optimum operating point 93-94
    outrunner and inrunner 57
    positioning and mounting 16, 18, 20, 22-24, 101, 118, 133
    quantity 25
    redundancy 23
    reversing the direction 65-66
    sizes 25, 58, 91-93
    torque 21

National Museum of the USAF, Dayton, Ohio 9
Nuclear bomb testing 11

Off-the-shelf aircraft 16
Orthophotos 28-29

Paparazi UAV 12
PID tuning 49-50, 115, 145
Pitot tube 47
Power distribution boards (PDBs) 40-41, 106-108, 115, 119-120
Power module (current and voltage sensor) 46
Pre-flight safety checks 145-149
Professional filming drones 23
Project Aphrodite 11
Propellers – see also Rotor blades 17-18, 21, 24, 66-71, 100
    balancing 69-71
    checking 146
    choosing 91-92, 99
        fixed-wing 92-93, 124
        multirotor 91-92
    direction of spin 135
    fastening 68
    foldable 67
    materials 66
    prop adapters 68-69
    removing for safety 115, 139
    sizes (diameter and pitch) 25, 58, 67, 91-94, 150
Puller (tractor) configuration 17-18
Pusher configuration 16-19, 133

Racing 35, 91
Radio frequency (RF) transmissions 10, 51, 55-57
    PPM (pulse-position modulation) 56-57, 112, 121
    PWM (pulse width modulation) 56-57, 121
    2.4GHz 56
R/C (radio-controlled) aircraft 13, 19, 34, 39, 51-57, 61, 125, 144
R/C controllers 37-38, 47
    buddy box 148
R/C receivers 55, 76, 112, 116, 121, 125
    binding to transmitter 55
    connecting to ESC 135
    connecting to flight controller 121
    mounting on drone 137
    tuning tor transmitter 121

R/C transmitters 52-55, 115, 121, 139
   channels 53
   check communication with autopilot 147
   failsafe 54
Real-time (live) video 16, 28, 32, 34, 38, 115
Repairs and upgrades 16
Research and teaching 35
Roma, Italian warship 10
Rotor blades – see also Propellers 19, 25
   counter-rotating 19
   pitch angle 19
Rotorcraft 16, 19-25
   ducted fan copters 24-25
   helicopters 10, 19-20, 25
   hexacopters 19, 22-23
      ex (X) configuration 22
      plus (+) configuration 22
      Y6 configuration 22-24
   multicopters/multirotor 19-21, 23, 25, 28, 31, 33, 39, 91-92, 99, 145, 148
      arms 21-22, 40, 104, 119
      central plate 39-40, 113
   octocopter 22-24
      X8 configuration 24-25
   quadcopters 10, 19-21, 24, 28, 34-35, 37, 39, 41, 49, 60, 95, 100-115
      ex (X) configuration 21
      mini FPV 116-123, 149
      plus (+) configuration 21
      toy 145
   tricopters 20-21, 23, 147
   X-copters 24
      Bicopters 25
      SingleCopters 24-25
RPVs (remotely-piloted vehicles) 8
Ruston Proctor AT 9
Ryan Aeronautics Co 11

Safety features and aspects 25, 28, 33, 115, 139, 146, 148
   parachutes 146
Satellite imagery 29-30
Security drones 32-33
Scratch-building 16
Sensors 12-13, 34, 43-47
   calibration 45
   compass module (magnetometer) 45, 108, 111-112
      calibration 45
   distance 46
   lightweight gamma spectrometer 31
   optic flow 47
   radiation 31
   ultrasonic 46
Servos 18-20, 38-39, 43, 49, 147
   attaching to wings 127-131
   cables 129
   horn connectors 126

Skaug, Steffen (BLheli firmware) 61
Soldering 103-105, 114, 118, 134-135
Source codes 12-13, 35, 61
Sperry, Elmer 10
Sperry, Lawrence 9
Steam-powered aircraft 9
Surveillance drones 12, 32-33

Tail section – see Empennage
Target drones 10-12
TDR-1 assault drone 10-11
Teledyne Ryan Firefly/Firebee drones 11-12
Telemetry systems 51-52, 111, 138
   links 36, 38
Tesla, Nikola 9
Thermopiles 13
Thrust data tables 93-94, 124-125, 150-151
   how to read 150-151
   how to record 151
Twin-boom aircraft 18-19

UAVs (unmanned aerial vehicles) 8, 12, 17, 31
USB connections 62-63, 123
US Navy 10

Vibrations 49, 77, 101, 109, 137, 146
Video monitors 37-38
Video-piloting – see FPV
Video recording 76
Video transmitters 38, 52, 80-84
Vietnam War 12
Voltage regulators 40, 122

Websites
   community forum (Drone Trest) 139, 149
   Pix4D 29
Weight 98-100, 116, 124-125
   maximum take-off 25, 124-125
Windy conditions 20, 25, 47, 49, 51, 92
Wings 17, 39
   adding extra equipment 131
   building 126-132
   chord 125
   control surfaces 127-128
   foam 127
   loading 124
   span 125
   winglets 132
World War One 10
World War Two 10-11
Wright brothers 9

Yaw mechanism 20

3D printers 39